Public Health Ethics Analysis

Volume 8

Series Editor
Michael J. Selgelid, Centre for Human Bioethics, Monash University, Melbourne, VIC, Australia

During the 21st Century, Public Health Ethics has become one of the fastest growing subdisciplines of bioethics. This is the first Book Series dedicated to the topic of Public Health Ethics. It aims to fill a gap in the existing literature by providing thoroughgoing, book-length treatment of the most important topics in Public Health Ethics—which have otherwise, for the most part, only been partially and/or sporadically addressed in journal articles, book chapters, or sections of volumes concerned with Public Health Ethics. Books in the series will include coverage of central topics in Public Health Ethics from a plurality of disciplinary perspectives including: philosophy (e.g., both ethics and philosophy of science), political science, history, economics, sociology, anthropology, demographics, law, human rights, epidemiology, and other public health sciences. Blending analytically rigorous and empirically informed analyses, the series will address ethical issues associated with the concepts, goals, and methods of public health; individual (e.g., ordinary citizens' and public health workers') decision making and behaviour; and public policy. Inter alia, volumes in the series will be dedicated to topics including: health promotion; disease prevention; paternalism and coercive measures; infectious disease; chronic disease; obesity; smoking and tobacco control; genetics; the environment; public communication/trust; social determinants of health; human rights; and justice. A primary priority is to produce volumes on hitherto neglected topics such as ethical issues associated with public health research and surveillance; vaccination; tuberculosis; malaria; diarrheal disease; lower respiratory infections; drug resistance; chronic disease in developing countries; emergencies/disasters (including bioterrorism); and public health implications of climate change.

Susan Bull • Michael Parker • Joseph Ali
Monique Jonas • Vasantha Muthuswamy
Carla Saenz • Maxwell J. Smith • Teck Chuan Voo
Jantina de Vries • Katharine Wright
Editors

Research Ethics in Epidemics and Pandemics: A Casebook

 Springer

Editors
Susan Bull
The Ethox Centre, Nuffield Department
of Population Health
University of Oxford
Oxford, Oxfordshire, UK

Michael Parker
The Ethox Centre, Nuffield Department
of Population Health
University of Oxford
Oxford, Oxfordshire, UK

Joseph Ali
Johns Hopkins Berman Institute of Bioethics
Johns Hopkins Bloomberg School of Public
Health
Baltimore, MD, USA

Monique Jonas
School of Population Health
University of Auckland
Auckland, New Zealand

Carla Saenz
Regional Program on Bioethics
Pan American Health Organization (PAHO)
Washington, DC, WA, USA

Vasantha Muthuswamy
Forum for Ethics Review Committees in India
Parel, India

Maxwell J. Smith
Faculty of Health Sciences
Western University
London, ON, Canada

Teck Chuan Voo
Centre for Biomedical Ethics
National University of Singapore
Singapore, Singapore

Jantina de Vries
Department of Medicine
University of Cape Town
Cape Town, South Africa

Katharine Wright
Freelance Bioethics Consultant (formerly Nuffield
Council on Bioethics)
London, UK

ISSN 2211-6680   ISSN 2211-6699   (electronic)
Public Health Ethics Analysis
ISBN 978-3-031-41803-7   ISBN 978-3-031-41804-4   (eBook)
https://doi.org/10.1007/978-3-031-41804-4

This Springer imprint is published by the registered company Springer Nature Switzerland AG
The registered company address is: Gewerbestrasse 11, 6330 Cham, Switzerland

Paper in this product is recyclable.

Foreword

Katherine Littler

In response to the recognition that the nature and complexity of public health emergencies raise both profound and distinct challenges in terms of how to appropriately undertake health-related research, WHO's Health Ethics and Governance Unit put out a call for proposals to better understand the breadth of ethical issues associated with the COVID-19 pandemic. COVID-19 certainly demonstrated the ethical issues that could arise on an unprecedented global scale. The call for proposals was issued in the latter half of 2020, when the COVID-19 pandemic was still in its relatively "early stages" and there was, and still remains, a real need to better understand the breadth of ethical issues associated with the pandemic and also to consider appropriate ways of addressing them. While the work and projects that resulted from this call do not necessarily represent the viewpoints of the WHO, it was the intention that this work should help to inform the future work of the Health Ethics and Governance Unit, the Epidemic Ethics initiative and those working to ensure that ethics is a key component of both the COVID-19 response and future preparedness.

One of the truly global and inclusive responses to the call for proposals is this casebook. The casebook reflects the experiences of researchers and ethics committees from 5 of the 6 WHO regions, and the 44 case studies have a strong focus on low- and middle-income settings, and the challenges experienced when conducting research with some of the populations most adversely affected by COVID-19. The diversity and breadth of the issues captured reflect a richness of issues and subject matter, which include the following: justifying, prioritizing and adapting research in rapidly evolving pandemic contexts; pandemic research exceptionalism, quality and publication; adjusting ethical review; surveillance and sharing of individual-level health data; and responsibilities to participants and communities during research and rollout.

This casebook builds on the foundations laid by the 2009 WHO *Casebook on ethical issues in international health research*, edited by Richard Cash et al., which has demonstrated the enormous value of the casebook format in narrowing the gap between conceptual ethical analysis and real-world lived experiences, by promoting a deeper understanding of the practical ethical issues in ways that can be translated and integrated into both our research and our responses to public health emergencies. This casebook aims to serve as a valuable tool for present and future public health emergencies, supporting the breadth of people working in this area from both regional and technical perspectives; including researchers, research ethics committees, health authorities, engagement practitioners and publics, among others. Critically, the focus is on case studies reflecting experiences which may not usually be shared for discussion and consideration. This book is an invitation to discuss, deliberate on and understand ethical issues arising in pandemic research in order to build solidarity, equity and access to research for the future.

Preface

There is widespread recognition that the nature and complexity of epidemics and pandemics raise a number of complex and important ethical challenges for the conduct of health-related research. The importance of conducting robust research to effectively address knowledge gaps about how best to prevent and control infection caused by novel and evolving pathogen variants, and about addressing disease burdens effectively, is undeniable. However, the complexity and unpredictability of pandemics and epidemics, and the substantial strain they place on health systems and capacities to conduct research and provide effective oversight of it, raise a variety of profound ethical challenges.

The scale and pace of research into the diagnosis, treatment and prevention of health burdens in the context of the COVID-19 pandemic is unprecedented. Stakeholders involved in pandemic research pathways have highlighted a range of ethical issues requiring consideration, including issues associated with expedited research and review pathways; prioritizing and suspending research; pandemic exceptionalism; unprecedented rates of pre-publication, publication and retraction of research findings; responses to structural inequities and vulnerabilities; research quality and misconduct; adapting and adaptive research; and emergency use of unregistered and investigational interventions, among others. Given the range and complexity of issues that researchers and reviewers encounter in epidemics and pandemics, the development of training resources to support capacity strengthening in epidemic and pandemic research ethics is a priority to enable effective design, review, conduct and oversight of research.

There are currently very few practical training resources that focus on research ethics specifically in the context of epidemics and pandemics. This casebook is envisioned as a resource to assist researchers, research ethics committees and regulators to assess and promote the ethical conduct of research in epidemics and pandemics. Case studies are a highly adaptable capacity-building approach used within a range of disciplines to facilitate higher levels of cognitive engagement with analysis, evaluation and application of relevant concepts. We have found that in the context of health and research ethics, case studies play a key role in strengthening capacities for ethical analysis and promoting a deep understanding of relevant ethical issues, differing perspectives and competing considerations, as well as thoughtful evaluation of their implications for practice.

In early 2021, we issued a global call for submissions of real-world case studies of ethical issues arising during the design and implementation of research responses in epidemics and pandemics. This casebook brings together 44 cases from around the globe, with a specific focus on the COVID-19 pandemic. Each case is relatively short (up to 1000 words) and is based on an actual research project or context, and key ethical issues which it raised. Cases are accompanied by a small number of suggested questions, which highlight a range of complex, and at times inter-related, ethical considerations. As readers become familiar with the cases, it is likely that they may identify additional ethical issues of interest and questions which have not been explicitly addressed. We have not sought to answer the questions posted in the cases – as can be seen, often there is no single "correct" answer appropriate for all contexts in which these questions may arise. Instead, the aim is to promote consideration of the relevant ethical dimensions and evaluation and justification of contextually appropriate responses.

As the cases in this casebook are drawn from lived experiences, for almost all of the cases, potentially identifying details have been redacted and descriptions of specific contexts generalized. De-identification has been undertaken to promote a focus on considering the ethical dimensions, including how the issues raised in the case may be appropriately addressed in the reader's own setting, and to remove private and/or sensitive information. A very small number of cases that highlight specific and important issues could not be effectively de-identified, owing to their unique characteristics, and have been included with appropriate permissions.

The cases have been grouped into thematic chapters, based on core issues they raise in pandemic contexts. Each chapter commences with an introduction, which outlines relevant conceptual approaches and ethical considerations relating to each theme, referring to the cases which follow. The allocation of cases to chapters is intended to facilitate, not constrain, consideration of the practical ethical issues they address. Given that many cases prompt considerations of issues which are relevant to more than one chapter, a case keyword overview has been included, which indicates a range of cases it may be valuable to draw on when exploring specific ethical issues.

The chapters in this casebook reflect core themes arising in the case studies submitted, and do not follow a structure common in research ethics standards, where discrete topics, such as research justification, acceptable risks and burdens, and consent, are addressed sequentially. Chapter 1, "Introduction: Research Ethics and Health Policy in Epidemics and Pandemics", introduces the changing, and at times competing, obligations, values and priorities underpinning complex interactions between researchers, regulators and policy-makers during the design, conduct and modification of pandemic research, and early implementation of research findings, against a background of rapidly evolving public health policies and pandemic responses. Chapter 2, "Setting Research Priorities", outlines the ethical concepts and substantive and procedural considerations arising when making decisions about stopping, pausing or revising ongoing research and adapting research priorities and plans in response to epidemics and pandemics. Chapter 3, "Research Quality and Dissemination", reviews practical ethical issues and moral questions which arise when seeking to ensure that both the design and the conduct of research, as well as the dissemination and publication of research findings, are of appropriate quality, given the constraints inherent in epidemics and pandemics. Chapter 4, "Boundaries between Research, Surveillance and Monitored Emergency Use", addresses how boundaries between research activities and public health activities can rapidly blur and change in epidemics and pandemics, and the challenges that arise when seeking to ensure both that research activities can be clearly identified and thus meet appropriate ethical standards, and that research and public health activities can be effectively co-ordinated in pandemic responses. Chapter 5, "Adapting and Adaptive Research", highlights responsibilities to ensure that research is both appropriately flexible and responsive to rapidly evolving pandemic landscapes, the ethical considerations which inform decisions about when adaption is needed, and the importance of ensuring that research continues to meet ethical standards when adaption is necessary. Chapter 6, "Ethics Review Challenges", outlines the considerations and tensions which can arise when seeking to establish agile research ethics processes for rapid and streamlined review of priority studies, while also ensuring that ethical standards are maintained in burdensome pandemic contexts. Chapter 7, "Ethical Issues Associated with Managing and Sharing Individual-Level Health Data", explores some of the complex questions that can arise when seeking to maximize the utility of clinical, surveillance and research data to inform public health responses to the pandemic, while ensuring that the interests of data subjects are appropriately respected and public trust is maintained. Chapter 8, "Dimensions of Vulnerability", provides a conceptual overview of responsibilities to ensure that research is appropriately responsive to the inequitable impact of the COVID-19 pandemic. Chapter 9, "Participant Recruitment, Consent and Post-trial Access to Interventions", reviews the practical ethical challenges that can arise during the conduct of research, as researchers seek to fulfil obligations to respect the rights and interests of participants while also conducting robust research to address population health needs, and ensuring that research procedures are appropriately tailored to pandemic constraints. Chapter 10, "Afterword", reflects on the breadth of research required to inform prevention, preparedness, response and recovery from public health emergencies, and the resources that could complement this casebook.

The cases and materials in this book represent the opinions and conclusions of the authors, and do not necessarily reflect the views and policies of the editors, or the editors' or authors' institutions. We were honoured by the enthusiastic and thoughtful responses to the call for cases and materials for this casebook. It has been a pleasure to work with over 80 authors from Africa, the Americas, Asia, Europe and Oceania on the co-creation of this casebook, and we extend our thanks to all those whose experience and expertise have made this such a rich resource.

Susan Bull Associate Professor, The Ethox Centre and Wellcome Centre for Ethics and Humanities, Nuffield Department of Population Health, University of Oxford, and Associate Professor, Department of Psychological Medicine, Faculty of Medical and Health Sciences, University of Auckland

Michael Parker Director of the Wellcome Centre for Ethics and Humanities, and the Ethox Centre, Nuffield Department of Population Health, University of Oxford

Joseph Ali Associate Director for Global Programs, Johns Hopkins Berman Institute of Bioethics and Assistant Professor, Department of International Health, Johns Hopkins Bloomberg School of Public Health

Monique Jonas Associate Professor, School of Population Health, University of Auckland

Vasantha Muthuswamy Former Senior Deputy DG, ICMR (Indian Council of Medical Research) and President, FERCI (Forum for Ethics Review Committees in India) and Chair COVID19 National Ethics Committee (CoNEC)

Carla Saenz Regional Bioethics Advisor, Regional Program on Bioethics, Department of Evidence and Intelligence for Action in Health, Pan American Health Organization

Maxwell J. Smith Assistant Professor, Faculty of Health Sciences, Western University & Dalla Lana School of Public Health, University of Toronto.

Teck Chuan Voo Assistant Professor, National University of Singapore Centre for Biomedical Ethics (CBmE)

Jantina de Vries Associate Professor in Bioethics, University of Cape Town

Katharine Wright Freelance Bioethics Consultant (formerly Nuffield Council on Bioethics)

Oxford, UK and Auckland, New Zealand	Susan Bull
Oxford, UK	Michael Parker
Baltimore, MD, USA	Joseph Ali
Auckland, New Zealand	Monique Jonas
Parel, India	Vasantha Muthuswamy
Washington, DC, WA, USA	Carla Saenz
London, ON, Canada	Maxwell J. Smith
Singapore, Singapore	Teck Chuan Voo
Cape Town, South Africa	Jantina de Vries
London, UK	Katharine Wright

In Memoriam: Vasantha Muthuswamy

Vasantha in September 2022 in video meeting with the PREPARED project.

Fig. 1 Vasantha Muthuswamy

See Fig. 1.

> Our life is full of interpunctions, or commas; death is but the period or full point. (Thomas Jackson; Maran Atha; A. Maxey; 1657)

Born in Chennai (erstwhile Madras), Tamil Nadu, on 12 July 1948, Dr. Vasantha Muthuswamy's journey came to a full point on 21 February, 2023, leaving behind her son, daughter-in-law, two sisters and one brother.

After finishing school with flying colours and Pre-University at Chennai, her education continued for a Pre-professional course in Kolkata. She pursued a medicine (MBBS) course and a postgraduate Diploma in Obstetrics and Gynaecology at the R.G. Kar Medical College, Kolkata. She was a topper throughout her studies. Soon after qualifying for MD in 1979 as a Gold medallist from the Institute of Obstetrics and Gynaecology, Madras Medical College, Chennai, she was selected as a scholar in the Science Talent Scheme initiated by the visionary Dr. C. Gopalan, Director General, Indian Council of Medical Research (ICMR). After that training, she got a placement as an ICMR Scientist at the Toxaemia Research Unit, Vani Vilas Hospital, Bangalore in 1979 and then at the ICMR Institute, currently named the National Institute of Reproductive Research and Child Health, Mumbai in 1980. She was transferred to ICMR headquarters in 1982 where she worked for some time in the Division of Reproductive Health and Nutrition, the Indo-foreign cell and the Division of Basic Medical Sciences (BMS) before she went to Lal Bahadur Shastri National Academy of Administration, Mussoorie on a contract appointment. On her return, she was given charge of the Division of BMS as its Chief. In 2009, she became Senior Deputy Director General in the Division and made its several activities a strength to reckon with. Facing challenges with determination and confidence was second nature to her. Her approachable demeanour and leadership qualities were assets which made her junior scientists and staff give her support for efficiency. Later, she was given additional charge as Director-in-Charge of the current National Institute of Immuno-haematology (2003–2005) and 7 months before retirement as Head of the Division of Reproductive Health and Nutrition in 2008.

For revising ICMR's 1980 Policy Statement on Ethics under the Chairmanship of Hon'ble Chief Justice of India, Sh. M. N. Venkatchaliah, the training she received on a WHO Fellowship in Bioethics from the Kennedy Institute of Bioethics, George-town University, USA, became useful. Later she became a renowned national and international expert in that area. She was associated with the formulation of several ethical guidelines in India, the last one being on COVID-19. Other countries like Nepal, Sri Lanka and agencies like the WHO, UNAIDS, CIOMS, Nuffield Council, Family Health International and HIV Prevention Trials Network benefitted from her contribution. She was part of training for setting up ethics committees in Cambodia, Laos PDR and Maldives. She was the founder member secretary of the committee initiated by WHO TDR in formulating operational guidelines for ethics committees in 2000 which led her to become a member of the Steering Committee of the Forum for Ethics Review Committees of Asia Pacific Region (FERCAP). She was Co-investigator for NIH-funded bioethics education activities in India. She was associated with the Indian Journal of Medical Ethics as a member of its Editorial Advisory Board and also an active proponent of Ethics at the National Bioethics Conferences that were organised by the Journal. She was one of the editors of the first volume of 'Biomedical Ethics Perspectives in the Indian Context', an ICMR publication. As President of the Forum for Ethics Review committees in India, the National Chapter of FERCAP, she was associated with EU project 'TRUST' for the formulation of the 'Global Code for the Conduct of Research in Resource-poor Settings'. In the later few months of her life, she was engaged with another EU project 'PREPARED'. Dr. Vasantha was an Editor on this casebook, and her enthusiasm, wisdom and valuable contributions to the casebook from January 2021 until December 2022 were greatly appreciated by the editorial team.

She sowed the seeds of 'Research Ethics' in India which are bearing fruits now. She was a good teacher, orator and advisor in many other areas besides bioethics such as drug development, genetics, genomics, genetically modified food, haematological diseases, traditional medicine etc. Even after retirement, she was very active in executing her tasks. She was Chair and a member of many national and international committees and ethics committees. Her ailments were never a hindrance to pursuing a task entrusted to her because she had the confidence and grit to face such issues as trivialities. She will be lovingly remembered for her energy and passion, warm-heartedness, untiring attitude and ever-smiling face. Unfortunately, this daunting spirit ebbed away gradually in the last few months of her illness and finally, she gave in plunging her family, friends and colleagues into a sense of deep loss. The light has gone but her spirit lives with us. May she be at peace in her heavenly abode!

Case Study Authors

Juan Manuel Alba Bermúdez. Law School, Universidad de Las Américas, Quito, Ecuador

Melchor Alpízar-Salazar. Centro Especializado en Diabetes, Obesidad y Prevención de Enfermedades Cardiovasculares, Mexico & Instituto Nacional de Investigación Avanzada en Diabetes Mellitus y Enfermedades Crónicas del Metabolismo, Mexico

Dulce María Fernanda Alpízar-Sánchez. Centro Especializado en Diabetes, Obesidad y Prevención de Enfermedades Cardiovasculares, Mexico, Instituto Nacional de Investigación Avanzada en Diabetes Mellitus y Enfermedades Crónicas del Metabolismo, Mexico

Jennyfer Radeino Ambe. Department of Public Health, Koinadugu College, Kabala, Sierra Leone & The Coalition for Equitable Research in Low-Resource Settings (CERCLE), Paris, France

Action, Amos. Centre for Clinical Brain Sciences, University of Edinburgh, Edinburgh, United Kingdom & International Bureau of Epilepsy, Pan African Network for Persons with Psychosocial Disabilities

Verónica Anguita Mackay. Ethics Committee, Fundación Arturo López Pérez, Santiago, Chile, & Universidad de Chile, Chile

Femke Bannink Mbazzi. MRC/UVRI & LSHTM Uganda Research Unit, Entebbe, Uganda & Ghent University, Ghent, Belgium

Capucine Barcellona. SHAPES Initiative, Centre for Biomedical Ethics, Yong Loo Lin School of Medicine, National University of Singapore, Singapore,

Javiera Bellolio Avaria. Instituto de Estudios de la Sociedad, Santiago, Chile

Nikola Biller-Andorno. Institute of Biomedical Ethics and History of Medicine, University of Zurich, Zurich, Switzerland & Digital Society Initiative, University of Zurich, Zurich, Switzerland

Susan Bull. The Ethox Centre and Wellcome Centre for Ethics and Humanities, Nuffield Department of Population Health, University of Oxford, UK & Department of Psychological Medicine, Faculty of Medical and Health Sciences, University of Auckland, New Zealand

Dabota Yvonne Buowari. Department of Accident and Emergency, University of Port Harcourt Teaching Hospital, Port Harcourt, Nigeria

James J. Callery. Mahidol Oxford Tropical Medicine Research Unit, Faculty of Tropical Medicine, Mahidol University, Bangkok, Thailand & Centre for Tropical Medicine and Global Health, Nuffield Department of Medicine, University of Oxford, Oxford, UK

Michael H. Campbell. Faculty of Medical Sciences, The University of the West Indies, Bridgetown, Barbados

Jesica E. Candanedo P. Comité Nacional de Bioética de la Investigación (CNBI), Panamá & Ministerio de Salud, Departamento de Medicina Preventiva y Social, Facultad de Medicina, Universidad de Panamá

Sarah Carracedo. Regional Program on Bioethics, Pan American Health Organization (PAHO), Washington, DC, United States of America

Lynn M Chambonnet. Comité Nacional de Bioética de la Investigación (CNBI), Panama & Secretaría Nacional de Ciencia, Tecnología e Innovación (SENACYT), Panama

Dennis Chasweka. Department of Paediatrics and Child Health, KUHeS, Blantyre, Malawi

Phaik Yeong Cheah. Mahidol Oxford Tropical Medicine Research Unit, Faculty of Tropical Medicine, Mahidol University, Bangkok, Thailand & Centre for Tropical Medicine and Global Health, Nuffield Department of Medicine, University of Oxford, UK

Ng Chirk Jenn. SingHealth Polyclinics, Singapore, Duke-NUS Medical School, Singapore

Christopher Chiu. Department of Infectious Disease, Faculty of Medicine, Imperial College London, UK

Jesús Manuel De Aldecoa-Castillo. Centro Especializado en Diabetes, Obesidad y Prevención de Enfermedades Cardiovasculares, Mexico & Instituto Nacional de Investigación Avanzada en Diabetes Mellitus y Enfermedades Crónicas del Metabolismo, Mexico

Charalambos Dokos. Institute of Pharmacology, University Hospital of Cologne, Cologne, Germany

Donna M. Denno. Department of Pediatrics, University of Washington, Seattle, United States of America & Department of Global Health, University of Washington, Seattle, United States of America

Vanessa Jaelle Dor. Master in Global Health at Harvard Medical School, United States of America

José Alberto Galván-Magaña. Centro Especializado en Diabetes, Obesidad y Prevención de Enfermedades Cardiovasculares, Mexico & Instituto Nacional de Investigación Avanzada en Diabetes Mellitus y Enfermedades Crónicas del Metabolismo, Mexico

Raimundo Gazitúa Pepper. Hematology Unit, Fundación Arturo López Pérez, Santiago, Chile

Nithya Gogtay. Department of Clinical Pharmacology, KEM Hospital, Mumbai, India

María Inés Gómez. Centro de Bioética, Facultad de Medicina Clínica Alemana-Universidad del Desarrollo, Santiago, Chile

Nishakanthi Gopalan. Medical Humanities and Ethics Unit (MedHEU), Faculty of Medicine, Universiti Malaya, Kuala Lumpur, Malaysia

Katie Groom. Liggins Institute, University of Auckland, New Zealand & National Women's Health, Auckland City Hospital, New Zealand

Richard Haynes. MRC Population Health Research Unit, Nuffield Department of Population Health, University of Oxford, United Kingdom

Shakel Samara Shameeka Henson. Kingstown, St. Vincent and the Grenadines, West Indies

Peter Horby. Pandemic Sciences Institute, Nuffield Department of Medicine, University of Oxford, UK

Li Yang Hsu. Saw Swee Hock School of Public Health, National University of Singapore, Singapore & Yong Loo Lin School of Medicine, National University of Singapore, Singapore

Monique Jonas. School of Population Health, Faculty of Medical and Health Sciences, University of Auckland, New Zealand

Sharon Kaur. Centre for Law and Ethics in Science and Technology (CELEST), Faculty of Law, Universiti Malaya, Kuala Lumpur, Malaysia

Phang Kean Chang. Department of Pathology, Faculty of Medicine, University of Malaya, Kuala Lumpur, Malaysia & Medical Research Ethics Committee, University of Malaya Medical Centre, Kuala Lumpur, Malaysia

Rachel King. University of California San Francisco, San Francisco, United States & INSERM, University of Montpellier, CHU Montpellier, Montpellier, France

Katarzyna Klas. Research Ethics in Medicine Study Group (REMEDY), Faculty of Health Sciences, Jagiellonian University Medical College, Krakow, Poland

Nandini Kumar. Forum for Ethics Review Committees in India, KEM Hospital, Mumbai, India

Shuba Kumar. Samarth (Social Science Research NGO), Chennai, India

Markus Klaus Labude. SHAPES Initiative, Centre for Biomedical Ethics, Yong Loo Lin School of Medicine, National University of Singapore, Singapore

Juan Alberto Lecaros. ICIM, Faculty of Medicine, Clínica Alemana Universidad del Desarrollo. Santiago, Chile,

Timothy Nicholas Lee. SHAPES Initiative, Centre for Biomedical Ethics, Yong Loo Lin School of Medicine, National University of Singapore, Singapore

Ignacio Esteban León. Facultad de Ciencias Exactas, Universidad Nacional de La Plata, La Plata, Argentina & Consejo Nacional de Investigaciones Científicas y Técnicas (CONICET), La Plata, Argentina

Sergio Litewka. Institute for Bioethics and Health Policy, University of Miami Miller School of Medicine, Miami, FL, USA

Lorna Luco. Hospital Padre Hurtado, Santiago, Chile & Universidad de Santiago de Chile, Chile

Kyriakoula Manaridou. AMEOS Klinikum St. Josef Oberhausen, Germany

Tania Manríquez Roa. Institute of Biomedical Ethics and History of Medicine, University of Zurich, Switzerland & Digital Society Initiative, University of Zurich, Zurich, Switzerland

Maria Elena Marson. Facultad de Ciencias Exactas, Universidad Nacional de La Plata, La Plata, Argentina & Consejo Nacional de Investigaciones Científicas y Técnicas (CONICET), La Plata, Argentina

Guido Enrique Mastrantonio Garrido. Facultad de Ciencias Exactas, Universidad Nacional de La Plata, La Plata, Argentina & Consejo Nacional de Investigaciones Científicas y Técnicas (CONICET), La Plata, Argentina

Roli Mathur. ICMR Bioethics Unit, Indian Council of Medical Research & WHO Collaborating Centre for Strengthening Research Ethics in Biomedical and Health Research, Bengaluru, India

Helen McShane. Nuffield Department of Medicine, University of Oxford, United Kingdom

Christopher Moxon. Wellcome Centre for Integrative Parasitology, University of Glasgow, Glasgow, UK & Kamuzu University of Health Sciences (KUHeS), Blantyre, Malawi

Vasantha Muthuswamy. Forum for Ethics Review Committees in India, KEM Hospital, Mumbai, India

Chirk Jenn Ng. SingHealth Polyclinics, Singapore & DUKE-NUS Medical School, National University of Singapore, Singapore

Busisiwe Nkosi. Social Science Department, Africa Health Research Institute, KwaZulu-Natal, South Africa & School of Law, University of KwaZulu-Natal, South Africa

Deborah Nyirenda. Community Engagement & Bioethics, Malawi Liverpool Wellcome Trust, Blantyre, Malawi

Ana Palmero. Directorate of Research for Health, Ministry of Health, Buenos Aires, Argentina

Kathy Peri. School of Nursing Faculty of Medical and Health Science University of Auckland, New Zealand

Rafael Rodrigo da Silva Pimentel. School of Nursing, University of São Paulo, São Paulo, Brazil

Magalys Quintana T. Comité Nacional de Bioética de la Investigación (CNBI), Panama & Secretaría Nacional de Ciencia, Tecnología e Innovación (SENACYT), Panama

Elise Racine. The Ethox Centre and Wellcome Centre for Ethics and Humanities, Nuffield Department of Population Health, University of Oxford, United Kingdom & The Institute for Ethics in AI, University of Oxford, United Kingdom

Christina Reith. Nuffield Department of Medicine, University of Oxford, United Kingdom

José de Jesús Resendiz-Rojas. Centro Especializado en Diabetes, Obesidad y Prevención de Enfermedades Cardiovasculares, Mexico & Instituto Nacional de Investigación Avanzada en Diabetes Mellitus y Enfermedades Crónicas del Metabolismo, Mexico

Sofía P. Salas. Center for Bioethics, Faculty of Medicine, Clínica Alemana Universidad del Desarrollo. Santiago, Chile

Jennifer Salgueiro. National Institute of Infectious Diseases Evandro Chagas, Fiocruz, Rio de Janeiro, Brazil

Ana V. Sánchez U. Comité Nacional de Bioética de la Investigación (CNBI), Panama & Secretaría Nacional de Ciencia, Tecnología e Innovación (SENACYT), Panama

Marcelo José dos Santos. School of Nursing, University of São Paulo, Brazil

Edson Silva dos Santos. School of Nursing, University of São Paulo, Brazil

William K.H. Schilling. Mahidol Oxford Tropical Medicine Research Unit, Faculty of Tropical Medicine, Mahidol University, Bangkok, Thailand & Centre for Tropical Medicine and Global Health, Nuffield Department of Medicine, University of Oxford, UK

Janet Seeley. London School of Hygiene and Tropical Medicine, London, United Kingdom & Africa Health Research Institute, KwaZulu-Natal, South Africa & School of Nursing and Public Health, University of KwaZulu-Natal, South Africa & MRC/UVRI & LSHTM Uganda Research Unit, Entebbe, Uganda

Maryam Shahmanesh. Institute for Global Health, University College London, London, UK & Clinical Science Department, Africa Health Research Institute, KwaZulu-Natal, South Africa

Maxwell J. Smith. Faculty of Health Sciences, Western University, Canada & Dalla Lana School of Public Health, University of Toronto, Canada

Miranda Zoe Smith. Department of Infectious Diseases, University of Melbourne at the Peter Doherty Institute for Infection and Immunity, Victoria, Australia,

Sarah Sullivan. College of Education and Health Sciences, Touro University California, Vallejo, USA

Leticia Suwedi-Kapesa. Malawi-Liverpool-Wellcome Clinical Research Programme, Blantyre, Malawi & Liverpool School of Tropical Medicine, International Public Health, Liverpool, United Kingdom

Tivyashinee Tivyashinee. Medical Humanities and Ethics Unit (MedHEU), Faculty of Medicine, Universiti Malaya, Kuala Lumpur, Malaysia

Claude Vergès. Comité Nacional de Bioética de la Investigación (CNBI), Panama & Departamento de Medicina Familiar, Facultad de Medicina, Universidad de Panamá, Panamá

Marcin Waligora. Research Ethics in Medicine Study Group (REMEDY), Faculty of Health Sciences, Jagiellonian University Medical College, Krakow, Poland

James A Watson. Mahidol Oxford Tropical Medicine Research Unit, Faculty of Tropical Medicine, Mahidol University, Bangkok, Thailand, & Centre for Tropical Medicine and Global Health, Nuffield Department of Medicine, University of Oxford, UK

Jane Williams. Australian Centre for Health Engagement, Evidence and Values (ACHEEV), University of Wollongong, Australia

Michelle Wilson. Medical Oncologist Cancer and Blood Services, Auckland City Hospital, Auckland, New Zealand & Honorary Senior Lecturer Faculty of Medical and Health Sciences University of Auckland, New Zealand

Katharine Wright. Freelance Bioethics Consultant & former Assistant Director, Nuffield Council on Bioethics
Vicki Xafis. SHAPES Initiative, Centre for Biomedical Ethics, Yong Loo Lin School of Medicine, National University of Singapore, Singapore
Argentina Ying B. Comité Nacional de Bioética de la Investigación (CNBI), Panama & Departamento de Microbiología Humana, Universidad de Panamá, Panamá
Thembelihle Zuma. Social Science Department, Africa Health Research Institute, KwaZulu-Natal, South Africa & University of KwaZulu-Natal, South Africa

Case Keyword Overview

The 44 cases within this casebook have been allocated to thematic chapters based on key areas of focus. However the breadth, complexity, and at times inter-relatedness of the ethical considerations arising in the majority of cases prompt consideration of a broader range of ethical issues than those highlighted in specific chapters. The allocation of cases to chapters is intended to facilitate, not constrain, consideration of the practical ethical issues they address, and an overview of case keywords is provided below to indicate a range of cases it may be valuable to draw on when exploring specific ethical issues.

Chapter 1

Case 1.1: A Study on the Telemonitoring of COVID-19 Patients at Home

Keywords Boundaries between research, surveillance and rollout; Researcher roles and responsibilities; Resource allocation; Safety and participant protection; Data protection, access and sharing; Digital and remote healthcare and research

Case 1.2: Trial Unblinding Following Emergency Authorization of Vaccines

Keywords Boundaries between research, surveillance and rollout; Research design and adaption; Access to experimental treatments; Regulatory review; Researcher roles and responsibilities; Emergency Use Authorisation; Vaccines; Placebo control

Case 1.3: COVID-19 Controlled Human Infection Studies

Keywords Social and scientific value; Risk/benefit analysis; Safety and participant protection; Ethical review; Community engagement and participatory processes; Controlled human infection studies

Case 1.4: Early-Stage Investigations into Infectious Diseases

Keywords Boundaries between research, surveillance and rollout; Data protection, access and sharing; Research design and adaption; Researcher roles and responsibilities; Consent; Sample access and sharing

Chapter 2

Case 2.1: Should Death and Grieving During the Pandemic Be Studied?

Keywords Research priority setting; Risk/benefit analysis; Safety and participant protection; Vulnerability and inclusion; Qualitative research

Case 2.2: Should Widespread Off-Label Use of Medication Influence Research Prioritisation?

Keywords Research priority setting; Social and scientific value; Resource allocation; Treatment repurposing

Case 2.3: Studying the Treatment of COVID-19 Patients with Traditional Medicine

Keywords Research priority setting; Social and scientific value; Researcher roles and responsibilities; Ethical review; Traditional medicine

Case 2.4: Research Reprioritization During the COVID-19 Pandemic

Keywords Research priority setting; Social and scientific value; Research design and adaption; Vaccines

Case 2.5: Challenges with Continuing Cancer Research in a Publicly Funded Hospital

Keywords Research priority setting; Resource allocation; Access to experimental treatments; Non-COVID-19 research

Chapter 3

Case 3.1: Self-Experimentation in the Development of COVID-19 Vaccines

Keywords Researcher roles and responsibilities; Community engagement and participatory processes; Regulatory review; Research publication ethics; Vaccines; Citizen science; Researcher safety

Case 3.2: Research with Chlorine Dioxide in a Prison During the COVID-19 Pandemic

Keywords Researcher roles and responsibilities; Research misconduct; Vulnerability and inclusion; Safety and participant protection; Social and scientific value; Risk/benefit analysis; Ethical review

Case 3.3: Evaluating the Role of the BCG Vaccine as a Prophylactic in Elderly Populations

Keywords Vulnerability and inclusion; Ethical review; Consent; Social and scientific value; Vaccine repurposing

Case 3.4: Publication, Pre-publication and Retraction of Research: How a Pandemic Magnifies Concerns About Publication Ethics

Keywords Research publication ethics; Research misconduct; Researcher roles and responsibilities; Pre-prints; Retractions

Case 3.5: Retracted Research: Impacts and Outcomes

Keywords Research publication ethics; Research misconduct; Researcher roles and responsibilities; Ethical review; Regulatory review; Risk/benefit analysis; Data protection, access and sharing; Treatment repurposing; Multi-centre research; Retractions

Chapter 4

Case 4.1: Use of Convalescent Plasma in Severely Ill COVID-19 Patients

Keywords Boundaries between research, surveillance and clinical care; Data protection, access and sharing; Consent; Treatment repurposing; Emergency Use Authorisation

Case 4.2: COVID-19 Antibody-Testing Initiatives in a European Country

Keywords Boundaries between research, surveillance and clinical care; Researcher roles and responsibilities; Return of results

Case 4.3: Competing Priorities Under Pressure: Government Collaboration with Academic Institutions

Keywords Boundaries between research, surveillance and clinical care; Researcher roles and responsibilities; Ethical review; Safety and participant protection; Consent; Privacy and confidentiality

Case 4.4: Vaccine Research or Rollout?

Keywords Boundaries between research, surveillance and clinical care; Research design and adaption; Risk/benefit analysis; Vulnerability and inclusion; Vaccines

Chapter 5

Case 5.1: Adapting Face-to-Face Interviews to Respect Infection Control Measures

Keywords Research design and adaption; Privacy and confidentiality; Data protection, access and sharing; Vulnerability and inclusion; Qualitative research; Digital and remote healthcare and research; Researcher safety

Case 5.2: A Community-Based Intervention for Indigenous Older Persons with Mild to Moderate Dementia

Keywords Research design and adaption; Risk/benefit analysis; Pausing and halting research; Vulnerability and inclusion; Community engagement and participatory processes; Resource allocation; Non-COVID-19 research

Case 5.3: Suspending Participation in Research

Keywords Research design and adaption; Safety and participant protection; Risk/benefit analysis; Researcher roles and responsibilities; Access to experimental treatments; Vulnerability and inclusion; Digital and remote healthcare and research; Non-COVID-19 research

Case 5.4: Ethics and Adaptive Trials in the COVID-19 Pandemic

Keywords Research design and adaption; Vulnerability and inclusion; Consent; Ethical review; Researcher roles and responsibilities; Research priority setting; Risk/benefit analysis; Research publication ethics; Treatment repurposing; Multi-centre research; Pre-prints

Case 5.5: The Impact of New Scientific Evidence on On-going COVID-19 Studies

Keywords Research design and adaption; Pausing and halting research; Ethics committee remits and responsibilities; Ethical review; Researcher roles and responsibilities; Social and scientific value; Treatment repurposing

Chapter 6

Case 6.1: Ethics Approval of a Multi-centre Study: To Expedite or Not?

Keywords Ethical review; Ethics committee remits and responsibilities; Risk/benefit analysis; Data protection, access and sharing; Safety and participant protection; Research priority setting; Multi-centre research; Treatment repurposing

Case 6.2: Ethics Review of Multi-centre Trials: Challenges and Unforeseen Issues

Keywords Ethical review; Ethics committee remits and responsibilities; Safety and participant protection; Multi-centre research; Treatment repurposing

Case 6.3: The Importance of Effective Research Ethics Review

Keywords Ethical review; Ethics committee remits and responsibilities; Safety and participant protection; Research quality; Social and scientific value; Multi-centre research; Treatment repurposing

Case 7.3: Ethical Conduct and Review of Research

Keywords Ethical review; Privacy and confidentiality; Ethics committee remits and responsibilities; Consent; Qualitative research

Case 7.4: Informed Consent and Data Protection in the Context of Increased Use of Information and Communication Technologies

Keywords Consent; Privacy and confidentiality; Data protection, access and sharing; Ethical review; Vulnerability and inclusion; Digital and remote healthcare and research; Qualitative research

Case 7.5: Research into COVID-19 and Cancer in Populous Low-Income Neighbourhoods

Keywords Vulnerability and inclusion; Data protection, access and sharing; Consent; Ethical review; Privacy and confidentiality; Boundaries between research, surveillance and clinical care; Non-COVID-19 research

Chapter 8

Case 8.1: Should Pregnant Women Be Included in COVID-19 Vaccine Trials?

Keywords Vulnerability and inclusion; Risk/benefit analysis; Social and scientific value; Safety and participant protection; Placebo control; Vaccines

Case 8.2: Ethics and Research Policy in a Forensic Psychiatric Hospital

Keywords Vulnerability and inclusion; Safety and participant protection; Research design and adaption; Researcher roles and responsibilities; Privacy and confidentiality; Data protection, access and sharing; Risk/benefit analysis; Boundaries between research, surveillance and clinical care; Non-COVID-19 research; Digital and remote healthcare and research

Case 8.3: Studying the Impact of COVID-19 on Vulnerable Populations

Keywords Vulnerability and inclusion; Researcher roles and responsibilities; Resource allocation; Boundaries between research, surveillance and clinical care; Qualitative research; Digital and remote healthcare and research

Case 8.4: Inclusion of Persons with Disabilities in COVID-19 Research

Keywords Vulnerability and inclusion; Researcher roles and responsibilities; Social and scientific value; Community engagement and participatory processes; Resource allocation; Digital and remote healthcare and research

Chapter 9

Case 9.1: Ethical Challenges Arising When Recruiting Adolescent Minors by Telephone

Keywords Consent; Vulnerability and inclusion; Privacy and confidentiality; Safety and participant protection; Researcher roles and responsibilities; Research design and adaption; Qualitative research; Non COVID-19 research; Digital and remote healthcare and research

Case 9.2: Quantitative and Qualitative Research into Attitudes Towards COVID-19

Keywords Consent; Research design and adaption; Safety and participant protection; Risk/benefit analysis; Community engagement and participatory processes; Research priority setting; Qualitative research; Researcher safety; Digital and remote healthcare and research

Case 9.3: Seeking Consent to Research Involving the Use of Convalescent Plasma from COVID-19 Donors in the Treatment of Cancer Patients

Keywords Consent; Risk/benefit analysis; Researcher roles and responsibilities; Digital and remote healthcare and research

Case 9.4: A Study Involving Minimally Invasive Tissue Sampling in Adults Who Died from COVID-19

Keywords Consent; Researcher roles and responsibilities; Research design and adaption; Safety and participant protection; Researcher safety

Case 9.5: COVID-19 Clinical Trials: Placebo Group Participants and the Right to Access the Experimental Product

Keywords Access to experimental treatments; Resource allocation; Regulatory review; Vulnerability and inclusion; Risk/benefit analysis; Post-trial follow-up and monitoring; Placebo control; Vaccines

Learning and Teaching Guide

This casebook seeks to promote awareness and understanding of the breadth, complexity, and at times, inter-relatedness of ethical issues arising when conducting research within complex and rapidly evolving pandemic contexts. We hope it provides a useful resource to facilitate discussion, debate and learning about practical decision-making and approaches to resolving ethical challenges when research is conducted in such contexts. This casebook aims to support a range of capacity building approaches, including discussion groups, workshops, certificate courses and academic degree programmes, and to be of value to research ethics trainers and facilitators, and academics teaching about ethical theory and applied research ethics. As such it is intended to be accessible to a range of audiences with an interest in the ethical conduct of research, including those with limited previous experience with theoretical ethics and ethical analysis. Key potential audiences include researchers and front-line staff involved in the design and conduct of research, ethics review committee members and administrators, and students in research-related academic courses.

This casebook does not provide an introduction to the ethical theories and approaches that inform research ethics guidance and oversight more broadly, given the substantial introductory resources that are already available. Instead the chapter introductions draw on such approaches to provide an overview of conceptual approaches and policy frameworks to inform consideration of specific themes in research ethics and decision-making in complex real world pandemic contexts. When using these cases with audiences who have little familiarity with research ethics, trainers and facilitators may find it useful to suggest introductory research ethics resources that can inform case discussions. To complement the resources with which trainers and facilitators are already familiar, open access introductory courses addressing research ethics, and research ethics specifically in emergencies and epidemics, are available (Nuffield Council on Bioethics 2020; World Health Organization 2015; World Health Organization 2023). We have also developed an online resource with additional materials on research ethics on pandemics to support learning and teaching, which is available at https://epidemicethics.tghn.org/research-ethics-epidemics-and-pandemics-casebook-supplementary-resources/.

Cases

The 44 cases within this casebook provide contextually rich examples of ethical issues arising as health research was conducted to address pandemic burdens during 2020 and early 2021. They have been kept relatively concise (up to 1000 words) in order to prompt consideration of key ethical issues arising when research is conducted within challenging pandemic contexts. The cases have been allocated to thematic chapters based on key areas of focus. However, the majority of cases prompt consideration of a broader range of ethical issues than those highlighted in specific chapters, and many could have appropriately been included in other chapters. Moreover, readers are likely to identify additional ethical issues of interest and questions which have not been explicitly addressed in the cases. An overview of case keywords has been provided above, but as facilitators and learners become more familiar with cases, they may identify additional keywords and themes that are important and relevant to them. The allocation of cases to thematic chapters, and of keywords to cases, is intended to promote, rather than constrain, discussion and deliberation.

The nine thematic chapters in this casebook each contain between four and seven case studies that outline ethical issues which have arisen in practice. Each case is accompanied by three or four questions which seek to prompt reflection about ethical considerations, and to facilitate discussion. Each case was developed with authors who responded to a call for submissions. The editorial team worked with the authors to ensure that cases contained enough contextual detail to facilitate consideration of practical ethical issues. Information that could be considered identifying, private, or sensitive was excluded. Two cases that highlight specific and important issues could not be effectively de-identified, owing to their unique characteristics, and have been included with appropriate permissions.

The case studies are designed to encourage users to draw on, critically reflect on, and be able to discuss and consider justifications for their approaches to addressing the ethical challenges raised. Often there is no single "correct" answer that is appropriate for all contexts in which the questions posed in the case study may arise. Instead, careful consideration of relevant ethical dimensions and evaluation of contextually-appropriate responses is required.

Approaches to Using the Cases

In courses and training sessions, facilitators of case study discussions will draw on their expertise and experience to promote effective learning. Below we offer some ideas and suggestions for facilitators with less experience of case-based learning approaches, and for casebook readers undertaking independent learning.

Case Selection and Session Design

These cases and the accompanying chapter introductions can be used in a range of ways in differing learning contexts. In facilitator-led and academic courses and trainings, it can be valuable to focus on specific chapters, drawing on the references and supplementary resources to deepen understanding of the themes of most relevance to specific audiences and contexts (see https://epidemicethics.tghn.org/research-ethics-epidemics-and-pandemics-casebook-supplementary-resources/).

Conversely, the cases can be used as stand-alone resources, and drawn on independently to meet learning objectives. Given the complex and inter-related ethical issues arising in these real world cases, facilitators may find it useful to familiarise themselves with all the cases, and review the overview of case keywords, and bring together cases from multiple chapters to explore specific areas of interest. Casebook readers undertaking self-study may likewise wish to focus on topics of specific interest by starting with the thematic chapter introductions – or to focus on individual cases that highlight issues of particular relevance and importance. We also invite facilitators to adapt the details of cases to make them more relevant to specific contexts and learning outcomes where this would be helpful. For example, details of cases and questions can be revised, or further information relevant to a specific research context could be added. Some cases include details about the regulatory and policy settings, the political landscape, and/or the organisation of the health system in which the case arose. For some purposes, it will be useful to consider the extent to which ethical analysis of a case is dependent on those factors. Discussions can be enriched by considering how a similar case might – and should – be responded to in other settings, and why. Facilitators should select cases with attention to the sensitivities that they may raise for a given group, and reflect upon how to elicit contributions to discussions in safe and constructive ways.

Where possible, it is useful to provide learners with cases in advance, so they have an opportunity to read and reflect on them prior to discussion. It may also be valuable to ask learners to focus on a specific aspect of the case, or undertake a short activity, such as preparing a response to a question in advance of the discussion. It is important to additionally provide learners with an overview to relevant conceptual approaches and substantive themes relevant to the case(s) being discussed. In addition to the relevant chapter introductions in the casebook, resources such as local guidance, examples of debates and practice, and relevant literature may be drawn on. These materials may be provided prior to case discussion (via a presentation or resource for preliminary reading) or drawn on during or after discussion. Each approach has different benefits – providing conceptual resources prior to the discussion may enable learners to identify ethical issues associated with cases more easily, and to discuss them with greater confidence and depth of consideration. Alternatively, learners may find it more interesting and accessible to commence with a case, and then engage with relevant ethical concepts and conceptual approaches that are of value in addressing the issues that arose in the discussion.

Engagement and Discussion

In case discussions, the facilitator's role is to promote active engagement with ethical dimensions of cases, and support learners to identify and develop their own critical analyses of key considerations. A good understanding of relevant areas is important for facilitators, so that they can identify key points emerging during discussions, irrespective of the unfamiliar terminology learners may use to discuss considerations that are new to them. These cases address complex real-world scenarios in which people may reasonably disagree about how the issues arising should be addressed in practice. It is important that facilitators do not suggest that there is a specific "correct" answer that participants should reach. Instead, learners should be encouraged to develop their own critical analysis and justifications for the position they have taken. In some situations, facilitators may wish to guide discussants to initially address a specific issue within a case study, and then, once the discussion has reached an appropriate depth of analysis, move to the next consideration. Alternatively, learners may be encouraged to reflect on a case and discuss all of the ethical considerations they think are relevant. These can then be noted and addressed sequentially in discussion.

In preparing for discussion, it is valuable to reflect on the size of the learner group. Effective approaches often combine smaller group discussions of cases (with four to eight participants) followed by feedback to the whole cohort, and further discussion. Such approaches enable those who feel more comfortable exploring ideas in a more private setting to have an opportunity to contribute their views and explore differing perspectives in a smaller group, before hearing about and engaging with a range of views across the group as a whole. When working with groups with members who do not know each other, and particularly where members may be less confident in contributing to shared discussions, it can be useful ask groups to assign roles to members (e.g. facilitator, note-taker, spokesperson) to assist each member to feel secure in making a contribution. It may also be useful to note that everyone should have the opportunity to express themselves in small group discussions.

It is important to advise learners that disagreements may arise and can provide a valuable opportunity to reflect on why people may have differing perspectives, and to reflect on their own views in light of these. It may also be helpful to note that discussants may wish to voice opinions that they do not necessarily agree with, and may be controversial, but that they think that are nonetheless important to raise and consider. Learners should be reminded that any disagreements should be constructive, respectful and welcomed as opportunities to deepen ethical reflection.

As small group discussions progress, facilitators can circulate amongst groups to see if there are queries, and ask questions or make suggestions if discussions appear to have stalled. For example, facilitators may wish to prompt consideration of:

- What further information they may need to develop a more definitive response to questions posed, and why such information is needed.

- What modifications to research prioritisation, research design, ethics/regulatory review, and research conduct are permissible, desirable, or necessary to implement in pandemic research contexts.
- What aspects of pandemics make these modifications acceptable or necessary.
- The extent to which such modifications could valuably be implemented more widely, including in non-pandemic contexts.
- How aspects of the social, political, regulatory, and health system settings might determine how ethical challenges in pandemic research should be managed.

As the case discussions progress to whole group discussions, facilitators may find it useful to provide overviews of the key points addressed and insights arising, to consolidate learning and move discussions forward. When working with larger groups, asking members to raise a hand to indicate their initial response to an ethical question that a case raises can prompt discussion. Sometimes people have a sense of their position before they are able to explain their reasons for holding it. Informal polls of this type give an impression of the range of views within the group and provide a starting point for collaborative ethical reasoning. When there is reticence to present personal positions, facilitators can invite the group to propose reasons that might support a given position, without expressing personal endorsement.

Where discussions have revealed strong differences of opinion, it can be constructive to conclude by noting all the points of agreement as well as the areas of disagreement. Members could be asked to share or note down a point on which their ethical thinking changed or they encountered a challenge and/or a point that they will continue to reflect upon or remain uncertain of. These techniques can encourage continued engagement, a sense of the potential for effective collaboration across ethical divides, and reinforce the idea that ethical reflection is continual process.

References

Nuffield Council on Bioethics. 2020. *Research in global health emergencies: Ethical issues.* The Global Health Network. https://globalhealthtrainingcentre.tghn.org/research-global-health-emergencies-ethical-issues/

World Health Organization. 2015. *Ethics in epidemics, emergencies and disasters: Research, surveillance and patient care' and accompanying presentations.* The Global Health Network. https://globalhealthtrainingcentre.tghn.org/research-ethics-epidemics-pandemics-and-disaster-situations/

———. 2023. *Research ethics online training.* The Global Health Network. https://globalhealthtrainingcentre.tghn.org/research-ethics-online-training-v2/

Acknowledgments

This casebook is the result of a collaborative effort as part of an Epidemic Ethics/ WHO initiative. It has been undertaken with the generous support of the UK Department for International Development and Wellcome (Grants 214711/Z/18/Z, 221559/Z/20/Z and 096527/Z/11/Z).

We are grateful for the additional financial support for the development of specific cases and chapter introductions provided by: Centers for Disease Control; National Health and Medical Research Council Centre of Research Excellence (NHMRC CRE), the Australian Partnership for Preparedness Research on Infectious Disease Emergencies (APPRISE, NHMRC ID 1116530); The National Cancer Institute (Argentina); National Science Center, Poland, UMO 2020/01/0/HS1/ 00024; Singapore Ministry of Health's National Medical Research Council under its NMRC Funding Initiative grant (NMRC/CBME/2016).

Our grateful thanks to the many people who have contributed their valuable time and expertise during the development of this casebook. Isabel Tucker undertook nuanced and meticulous copy editing of the casebook. Kevin Childs supported casebook project management, including compiling and managing the affiliations of case study authors and developing a press strategy and resources. Halina Suwalowska assisted with developing the case keyword overview.

Disclaimer

The chapter introductions and cases in this book represent the experiences, opinions, and conclusions of the authors. They do not necessarily reflect the official position, views, or policies of the editors, the editors' host institutions, or the authors' host institutions.

Contents

Editors and Contributors

About the Editors

Susan Bull (BSc, LLB, MA, PhD) is an Associate Professor in Bioethics at the Ethox Centre, University of Oxford, United Kingdom, and an Associate Professor of Medical Ethics at the Faculty of Medical and Health Sciences, University of Auckland New Zealand. Her research interests centre on ethical dimensions of health and global health with a thematic focus on the exercise of epistemic power. Her conceptual and empirical research has addressed global health data sharing, consent to research, ethical review of research, controlled human infection (challenge) studies, and infectious disease outbreaks, epidemics and pandemics. Susan leads Epidemic Ethics, a global community of bioethicists and stakeholders involved in public health and research responses to public health emergencies. Susan has served in multiple advisory roles including as a member of the WHO Working Group for Guidance on Human Challenge Studies in COVID-19 and lead writer of WHO Guidance on the Ethical Conduct of Controlled Human Infection Studies. Susan has chaired and served on research ethics committees in the UK and New Zealand, and provided training to research ethics committees in Europe, Asia, Africa and New Zealand. In collaboration with colleagues at the University of Oxford, The Global Health Network, World Health Organization, Nuffield Council on Bioethics and Multi-Regional Clinical Trials Center at Brigham and Women's Hospital and Harvard, Susan developed a suite of free online research ethics training courses taken by over 450,000 learners.

Michael Parker is a Professor of Bioethics and Director of the Ethox Centre at the University of Oxford. Ethox is a multidisciplinary bioethics research centre with a major programme of research on global health bioethics. In 2012, together with colleagues in Kenya, Malawi, South Africa, Thailand and Vietnam, Michael established the Global Health Bioethics Network (GHBN). Since that time, GHBN has expanded to include colleagues in a number of other countries including: Brazil, Cambodia, Ghana, Laos, Malaysia, Myanmar and Nepal. Michael has a strong research interest in infectious diseases ethics. In 2019, together with Jeff Kahn at the Berman Institute of Bioethics, he established the Oxford-Johns Hopkins Global Infectious Diseases Collaborative (GLIDE) which conducts collaborative research on infectious disease ethics together with partners in a range of low- and middle-income countries. In addition to, and complementary to, these research interests, Michael has also been involved in a range of policy related activities in the arena of global health. From 2018 to 2020, he chaired a Nuffield Council of Bioethics international working group on the ethics of research in global health emergencies. From 2020 to 2022, he was a participant in the UK Government's Scientific Advisory Group for Emergencies (SAGE) which advised on COVID-19. And since 2020, he has been a member of the WHO COVID-19 Ethics & Governance Working Group.

Joseph Ali (JD) is an Associate Professor of International Health at the Johns Hopkins Bloomberg School of Public Health and Core Faculty/Associate Director for Global Programs at the Johns Hopkins Berman Institute of Bioethics. His research engages a range of challenges in global health ethics, and includes empirical and normative work to address the implications of emerging global mobile and digital technologies as applied in the context of health research, public health programs and disease surveillance. At Johns Hopkins, he serves as a member of the Bloomberg School of Public Health (JHSPH) Institutional Review Board (IRB) and has been involved in establishing and operating US National Institutes of Health (NIH) funded non-degree, master's, doctoral and post-doctoral training programs in bioethics at Johns Hopkins and with partners in Uganda, Ethiopia, Zambia, Botswana and Malaysia. He is Associate Editor for the *Journal of Empirical Research on Human Research Ethics* and teaches courses in international research ethics, public health ethics and digital health ethics at Johns Hopkins.

Monique Jonas is an ethical theorist based at the School of Population Health at the University of Auckland, New Zealand, where she teaches ethics to medical and health sciences students. She has a Doctorate in Medical Ethics from Kings College London. Her programme of research spans a wide range of ethical and political-philosophical concerns connected with health, including the ethics of advising; ethical aspects of parenting and decision making for children and young people; the relationship between the family and the state; research involving secondary data and population health. Monique is the Chair of the Health Research Council Ethics Committee and has served on the National Ethics Advisory Committee and the National Health Committee in New Zealand.

Carla Saenz is the Regional Bioethics Advisor at the Pan American Health Organization (PAHO), which serves as the World's Health Organization's Regional Office for the Americas. She is responsible for PAHO's Regional Program on Bioethics, which provides support on bioethics to countries in Latin America and the Caribbean, e.g. strengthening national research ethics systems, integrating ethics in health-related work, and building capacity in bioethics. Dr. Saenz also manages

PAHO's ethics review committee, which reviews research conducted with PAHO's involvement in the region. An elected fellow of the Hastings Center, she has authored numerous publications on different areas of bioethics, coedited the book *Public Health Ethics: Cases Spanning the Globe* and contributed to several ethics guidance documents. She has been responsible for the development of PAHO's zika ethics guidance and numerous ethics guidance documents issued by PAHO during the COVID-19 pandemic, including Catalyzing Ethical Research in Emergencies. Ethics Guidance, Lessons Learned from the COVID-19 Pandemic, and Pending Agenda. Dr. Saenz serves on the board of the International Association of Bioethics and the Steering Committee of the Global Forum on Bioethics in Research. She holds a PhD in Philosophy from the University of Texas at Austin and, before joining PAHO, she was at the Department of Bioethics at Clinical Center of the National Institutes of Health (NIH), and in the faculty in the Philosophy Department at the University of North Carolina at Chapel Hill.

Maxwell J. Smith (PhD) is an Assistant Professor and Western Research Chair in Public Health Ethics in the Faculty of Health Sciences at Western University in London, Ontario, Canada. He also serves as an Associate Director of Western's Rotman Institute of Philosophy and has cross-appointments in Western's Department of Philosophy, Department of Epidemiology and Biostatistics, and Schulich Interfaculty Program in Public Health. His research is in the area of public health ethics, with a focus on infectious disease ethics and the demands that health equity and social justice place on governments and institutions to protect and promote the public's health. Professor Smith has served in a number of advisory roles to governments and health authorities, including as a member of Ontario's COVID-19 Vaccine Distribution Task Force, the Public Health Agency of Canada's Public Health Ethics Consultative Group and World Health Organization's COVID-19 Ethics and Governance Expert Working Group, and serves as a consultant to Epidemic Ethics.

Jantina de Vries is the Director of The Ethics Lab at the Neuroscience Institute and an Associate Professor of Bioethics in the Department of Medicine at the University of Cape Town. The Ethics Lab brings together a multidisciplinary group of scholars that foster transformative ethics scholarship that centres Africa as the context and driver for global health ethics. She leads two core grants that support that work: a Wellcome Trust Research Development Programme that seeks to articulate how knowledge from the African humanities could and should inform on the ethics of new and emerging health technologies; and an award from the Fogarty International Centre (NIH) award that aims to develop an MSc degree that teaches ethics from the South. With colleagues at the University of Ghana and elsewhere, she is also contributing to the development of a solidarity index for global health funders. Jantina's primary expertise is in the ethics of African genomics research. She was a member of the WHO Genome Editing Expert Advisory Committee and is currently a member of the Research Ethics Board of Médecins Sans Frontières and the

Steering Committee of the Global Forum for Bioethics in Research. Jantina obtained her DPhil through The Ethox Centre at the University of Oxford (2011), and undergraduate and postgraduate degrees in sociology at Wageningen University (2003). She has published over 120 articles in international peer-reviewed journals. Her work is funded by the Wellcome Trust, the European Commission, the National Institutes of Health and the National Research Foundation of South Africa.

Teck Chuan Voo is trained in Philosophy and Medical Jurisprudence, and he is an Assistant Professor at the Centre for Biomedical Ethics, Yong Loo Lin School of Medicine, National University of Singapore. He researches on healthcare ethics and the ethics of infectious disease control and prevention, with focus on emergency situations. He is on the editorial board of the journals *Asian Bioethics Review* and *Public Health Ethics*; and the Springer Nature book series on Philosophy and Medicine, Public Health and Health Policy Ethics and its companion series, the SpringerBriefs in Public Health and Health Policy Ethics. Teck Chuan is an appointed member of various ethics committees in Singapore, including the Bioethics Advisory Committee and the National Medical Ethics Committee. He is on the Steering Committee of the Global Forum on Bioethics in Research and the ethics advisory board for UNITE4TB. He has served the WHO in various capacities in the development of clinical guidelines, and ethics guidance relating to epidemics and public health emergencies. He sits on the Board of Directors for the International Association of Bioethics.

Katharine Wright is a freelance ethics consultant, working with a number of international organizations including the WHO Health Ethics and Governance Unit, the MRCT Center of Brigham and Women's Hospital and Harvard, and the Council of Europe. She is also a member of the BMJ Ethics Committee. Her background is in the intersection of bioethics with policy and practice, with a particular interest in issues of power, vulnerability and inclusion in health policy and research. From 2007 to 2022, she was an Assistant Director at the UK-based Nuffield Council on Bioethics, directing projects on the ethical aspects of a wide range of policy issues, including ageing and dementia, the inclusion of children in research, the conduct of research in global health emergencies, the donation of bodily materials and the promotion of cosmetic procedures. Before joining the Nuffield Council, she spent 9 years at the UK House of Commons, briefing MPs of all political parties on health issues, and then 4 years in the NHS, monitoring the effect of the Human Rights Act on health law in England. During her time at the House of Commons, she was also seconded to the English Department of Health to work on patient consent.

Contributors

Ilana Ambrogi is a family and social medicine physician by training and Senior Research Analyst at Anis – Institute of Bioethics, Brazil. She has a PhD in bioethics, applied ethics, and collective health from PPGBIOS/Fiocruz/ENSP (Brazil) and is currently in a postdoctoral at PPGBIOS/Fiocruz and Epidemic Ethics. She received an MD from Northwestern University Feinberg School of Medicine in Chicago, USA, and completed a family and social medicine residency at Montefiore Medical Center – University Hospital for Albert Einstein College of Medicine in the Bronx, NY, USA. She earned a double BS in Biology and Psychology from the University Of North Carolina – Chapel Hill, USA. She is part of the Editorial Board of the journal *Developing World Bioethics*. Her current research interests are reproductive and sexual health and rights, gender equity, feminism, and bioethics in public health emergencies.

Luciana Brito is a clinical psychologist by training and the Co-director of the Anis – Institute of Bioethics, Brazil. At the University of Brasília, Brazil (UnB) she earned a master's in Clinical Psychology and Culture and a PhD in Health Sciences, Bioethics, Mental Health, and Human Rights. She is a coordinating member of ABRASCO's Bioethics working group. She is dedicated to social science, human rights, and bioethics research. Her current research interests are in ethics during epidemics, bioethics in health sciences and public health, and gender studies.

Sarah Carracedo holds a Lawyer Degree from the Pontifical Catholic University of Peru and a Master of Bioethics from Monash University (Melbourne, Australia). Since 2020, she has been working as a consultant of the Regional Program on Bioethics of the Pan American Health Organization (PAHO). She has been an IRB intern in the CC Department of Bioethics of the National Institutes of Health (Bethesda, USA) and has worked in the Research Ethics Unit of the General Office of Research and Technology Transfer of the National Institute of Health of Peru (OGITT-INS). She has also served as a member and an administrator and has been in the Secretariat of different research ethics committees of Peru. She currently teaches in the Faculty of Law of the Pontifical Catholic University of Peru and in the Faculty of Medicine of Cayetano Heredia University of Peru, and is an Adjunct Assistant Professor in the Department of Bioethics of Clarkson University.

Phaik Yeong Cheah is a Professor of Global Health and Bioethicist at the University of Oxford. She is the Head of Bioethics and Engagement at the Bangkok-based Mahidol Oxford Tropical Medicine Research Unit (MORU) where she manages MORU's community and public engagement programme. Phaik Yeong's research focuses on ethical issues arising in research with underserved populations, in particular how to ethically involve children, migrants, refugees, and other vulnerable groups in research. Her other area of research is how to promote fair and equitable sharing of individual-level health research data, which includes how to ensure that data sharing and big data does not exacerbate existing inequalities between higher- and lower-income setting researchers. Phaik Yeong has been a member of the Steering Committee of the Global Forum on Bioethics in Research (GFBR) since 2016. She has been the chair of the COVID-19 Clinical Research Coalition Data Sharing Working Group since 2020.

Sharon Kaur is an Associate Professor at the Faculty of Law, Universiti Malaya, where she has designed modules on healthcare law and ethics for undergraduate as well as postgraduate students including a number of modules on the Masters of Health Research Ethics (MOHRE) at the Faculty of Medicine. Her research interests primarily revolve around global health bioethics. She is keen on developing her work on migrant health ethics, particularly in relation to health research; as well as issues relating to the intersection of bioethics, law, and public policy. During the Covid-19 pandemic, she was fortunate to be part of the formation of the Malaysian Bioethics Community, which responded to the unprecedented challenges of the pandemic and setting up Clinical Ethics Malaysia, which provides, among other things, a free online clinical ethics consultation service. She is a member of the WHO Covid 19 Ethics and Governance Working Group, the Steering Committee of the Global Forum for Bioethics in Research (GFBR), and the International Expert Network of the Global Infectious Disease Ethics Collaborative (GLIDE). She is working with colleagues on a Wellcome Trust funded project to set up the Southeast Asian Bioethics Network.

Sergio Litewka completed his medical degree at the University of Buenos Aires and his master's in Public Health at the University of El Salvador, also in Buenos Aires, Argentina. He is a faculty member in the University of Miami Department of Surgery and the Director of Global Bioethics at the University of Miami Institute of Bioethics, a World Health Organization (WHO) on Ethics and Health Policies. As Global Bioethics Director, Dr. Litewka's work focuses on the development of research and education activities with international governmental organizations, universities, and the private sector on human subject protection and the responsible conduct of research. From 2005 to 2018, he was the International Director for the Collaborative Institutional Training Initiative (CITI Program), a web-based initiative for research ethics and responsible conduct of research education. During 2011 he served as a member of the International Research Panel at the US Presidential Commission for the Study of Bioethics Issues.

Ignacio Mastroleo has a PhD in Philosophy from the University of Buenos Aires (UBA). He specializes in research ethics as well as ethics of medical innovation and the use of unproven interventions outside clinical trials during public health emergencies. He undertook his PhD, Postdoc, and early research career under Dr. Florencia Luna's supervision, a world-leading bioethicist from Argentina. He is a tenured researcher at the National Research Council of Argentina (CONICET), the Associate Director of the Program of Bioethics of FLACSO Argentina, and lectures undergraduates in ethics at UBA. He is also an internationally recognized scholar. In 2014, he was a Caroline Miles Scholar at the Ethox Centre, University of Oxford. He also won the Velasco Suarez Award for Excellence in Bioethics, an award for public health given by the Pan American Health Organization (PAHO), the regional office for the Americas of the World Health Organization (WHO). From September 2019 to June 2023, he served as member of the International Association of Bioethics (IAB) board of directors. From March 2020 to January 2022, he served as an ad honorem expert member for WHO in the Ethics and COVID-19 Working Group and, currently, in the COVID-19, Ethics & Governance Working Group. He was the leading writer of the WHO ethics guidelines on emergency use of unproven interventions outside research (MEURI) published in May 2022. Since June 2022, Dr. Mastroleo has been leading a Philosophy Research Program on AI for Health at the Collaborative Research Institute "Intelligent Oncology" (CRIION) in Freiburg, Germany.

Maru Mormina has an interdisciplinary background, having been trained in the sciences and the social sciences. Spanning ethics, political philosophy, social epistemology, and science and technology studies, Maru's research is concerned with the relationship between strategic ignorance and epistemic injustice, and how these might shape processes of scientific knowledge production and of knowledge use in evidence-based policy, particularly in public health. Maru has applied these conceptual tools to the study of global inequalities in knowledge production and their intersection with colonial and postcolonial structures, and more recently to questions regarding the use and non-use of expert knowledge in public policy during crises, especially public health crises.

Tom Obengo is a Post-doctoral Research Fellow at the Department of Medicine, Faculty of Health Sciences, University of Cape Town, where he focuses on Epidemic Ethics. His current research interests include epidemic ethics, genetic therapies for debilitating illnesses, the African Ubuntu ethics, the ethics of care and gender roles in the context of pandemics, healthcare in resource-scare regions, and the influence of religion on bioethics. He holds a Doctor of Philosophy (PhD) degree in Applied Ethics, specializing in Bioethics, from the University of Stellenbosch. His doctoral dissertation titled "A Utilitarian Assessment of the Relevance of Genetic Therapies for HIV/AIDS in Africa, with Special Reference to the Situation in Kenya" was carried out and completed in the Department of Philosophy at the University of Stellenbosch. Tom teaches Bioethics to medics and research scientists in the Department of Family Medicine at the Kabarak University, at the Kenya Medical Research Institute (KEMRI), and in the Department of Philosophy at the University of Stellenbosch, and to Christian service students at Moffat College. He is a member of the Bioethics Society of Kenya (BSK) and the Institutional Scientific and Ethical Review Committee (ISERC) at the Kijabe Hospital.

Ana Palmero holds a JD (equivalent) degree from the University of Buenos Aires, Argentina, and was trained in Bioethics at the Latin American Faculty of Social Sciences (FLACSO). She is currently a consultant for the Global Health Ethics &

Governance Unit at WHO and has served as a consultant for the Regional Program on Bioethics at the Pan American Health Organization (WHO's Office for America) for the development of guidance to strengthen research ethics systems in Latin America during the COVID-19 pandemic. She has been responsible for the Research Ethics Program of the Directorate of Health Research at the Ministry of Health, Argentina, from 2016 to 2023. She is an Adjunct Professor in the Bioethics Program of the Caribbean Research Ethics master's degree Program of the University of Clarkson, USA, and in the Bioethics Program of FLACSO, Argentina. She is a member of the Ethics Review Board of Médecins Sans Frontières (Doctors without Borders) and serves on the Steering Committee of the Global Forum on Bioethics in Research. In Argentina, she is the Chair of the Research Ethics Committee (REC) of the Ministry of Health and a member of the REC of Dr. Ramos Mejia General Hospital.

Mira L. Schneiders is a postdoctoral Social Science Research Fellow, based at the Socio-Ecological Health Research Unit, Institute of Tropical Medicine in Antwerp (ITM) in Belgium. As a qualitative global health researcher with an interest in improving health among marginalized and vulnerable populations, Mira's work has focused on a range of topics including HIV, ageing, global health ethics, and AMR. Mira's research has largely been conducted in Southeast Asia, where she currently leads the social science work package of a multidisciplinary project on antimicrobial resistance in Indonesia. Previously, Mira worked as a Postdoctoral Researcher within the Bioethics and Engagement Department, Mahidol Oxford Tropical Medicine Research Unit, within the Centre for Tropical Medicine and Global Health at the University of Oxford. Here, she coordinated a multicounty qualitative study on the lived experiences of COVID-19 and worked on various topics in global health research ethics. Mira holds a Bachelor of Psychology (Hons) from the University of Manchester, a Master of Global Health Science, and a DPhil from the University of Oxford. During her doctoral research, Mira examined the ethical and global health challenges arising in the context of rapid population ageing, focusing on skip-generation households in Cambodia. Previously, Mira spent some years working in global health public institutions, including as Technical Officer at the Department of Ageing and Lifecourse at the World Health Organization (WHO), and in the field of HIV prevention at the WHO and various international NGOs in Cambodia.

Sarah Sullivan is a Global Health professional, Academic and Paediatric Nurse Practitioner (PNP). Ms. Sullivan is currently working as an Associate Professor and Chair of the Global Health Concentration at Touro University California, College of Education and Health Sciences, Public Health program. Professor Sullivan's research focuses on research ethics, migrant health, human resources for health, and cervical cancer. Sarah Sullivan is a Fulbright Scholar and supports research ethics programs/institutional review boards (IRBs) in public and private universities across the Americas. Ms. Sullivan has over 25 years of global health experience working in many countries for organizations such as CARE International, USAID, and WHO and providing education and research support to universities. She also has considerable experience working in health clinics as a PNP in the San Francisco, California area serving low-income populations. Sarah Sullivan has a Master of Science in International Cross-Cultural Nursing and a Master of Public Health with a specialty in Maternal and Child Health (MCH) and International Health from the University of California, San Francisco and University of California, Berkeley.

Halina Suwalowska is a Research Fellow in Global Health Bioethics at the Ethox Centre, Wellcome Centre for Ethics and Humanities (WEH) at the University of Oxford. She collaborates with the Oxford-Johns Hopkins Global Infectious Disease Ethics Collaborative (GLIDE) and Epidemic Ethics (EE) network led by the World Health Organization. Her research focuses on ethical and social issues related to managing dead bodies during natural disasters and epidemics. Specifically, she investigates the challenges faced by frontline staff and "last responders" when caring for the deceased. Prior to her current role, Halina worked as a postdoctoral researcher with RECAP (Research Capacity Strengthening and Knowledge to Support Preparedness and Response to humanitarian crises and epidemics). RECAP is a partnership led by the London School of Hygiene and Tropical Medicine, involving universities in the United Kingdom, Sierra Leone, and Lebanon, as well as leading humanitarian NGOs. Halina holds a DPhil in Population Health from the Ethox Centre, which she completed in 2020. Her doctoral research primarily focused on the ethical and political aspects of implementing Minimally Invasive Autopsy (MIA) in low-income settings. Prior to her involvement with the Ethox Centre, she gained experience working at the Wellcome Trust in London. Halina is also a valued member of the Global Health Bioethics Network.

Abbreviations

ACT	Access to COVID-19 Tools
CHIS	Controlled human infection studies
CIOMS	Council for International Organizations of Medical Sciences
ECA	Emergency or conditional approval/authorization
EUA	Emergency use authorization
GDPR	General Data Protection Regulation (of the European Union)
HCQ	Hydroxychloroquine
MEURI	Monitored emergency use of unregistered and investigational interventions
NPIs	Non pharmaceutical interventions
NRAs	National Regulatory Authorities
NREC	National Research Ethics Committee
PAHO	Pan American Health Organization
PCSI	Preventive and compulsory social isolation
PPE	Personal protective equipment
RCT	Randomized controlled trials
RECs	Research ethics committees
TM	Traditional medicine
WHO	World Health Organization

Chapter 1
Introduction: Research Ethics and Health Policy in Epidemics and Pandemics

Michael Parker, Susan Bull ⓘ, and Katharine Wright

Abstract Global health emergencies such as the COVID-19 pandemic are contexts in which it is critical to draw upon learning from prior research and to conduct novel research to inform real-time decision-making and pandemic responses. While research is vitally important, however, emergencies are radically non-ideal contexts for its conduct, due to exceptional uncertainty, urgency, disruption, health needs, and strain on existing health systems, amongst other challenges. This generates novel ethical challenges and a broader conception of research ethics is necessary to effectively address the complexity of pandemic research contexts. Going beyond traditional approaches to research ethics centring on the design of specific studies, this broader conception requires consideration of fundamental questions relating to the exercise of power and influence throughout research pathways, and a broader attention to both salient ethical issues, and the ethical responsibilities of stakeholders. These include important questions about responsibilities to gather evidence and generate knowledge systematically during emergencies, to implement policy responses in ways that are amenable to evaluation, and even potential moral obligations to participate in research. In situations of heightened uncertainty, additional questions arise about what constitutes sufficient evidence to justify the development and implementation of policy responses, and the responsibilities of scientific and social science researchers involved in policy-making processes. The four cases in this chapter prompt reflection on evolving and at times competing values and

M. Parker (✉)
The Ethox Centre, Nuffield Department of Population Health, University of Oxford, Oxford, Oxfordshire, UK
e-mail: michael.parker@ethox.ox.ac.uk

S. Bull
The Ethox Centre, Nuffield Department of Population Health, University of Oxford, Oxford, Oxfordshire, UK

Faculty of Medical and Health Sciences, University of Auckland, Auckland, New Zealand

K. Wright
Freelance Bioethics Consultant (formerly Nuffield Council on Bioethics), London, UK

© PAHO and Editors 2024
S. Bull et al. (eds.), *Research Ethics in Epidemics and Pandemics: A Casebook*,
Public Health Ethics Analysis 8, https://doi.org/10.1007/978-3-031-41804-4_1

responsibilities of policy-makers, regulators, health authorities and researchers during the design and conduct of research, and proposed early implementation of research findings. These cases highlight issues arising when conducting research of national importance in a pandemic, where researchers are required to liaise with authorities responsible for pandemic responses and address complex ethical issues, including protecting the interests of participants and publics when tensions arise between prioritising the completion of research and accelerating the rollout of novel health interventions. This chapter invites reflection on the practical ethical implications of commitments to undertake research during emergencies, including the nature and scope of the relevant responsibilities of a range of stakeholders.

Keywords COVID-19 pandemic · Research ethics · Public health emergencies · Boundaries between research, surveillance and rollout · Researcher roles and responsibilities · Resource allocation · Safety and participant protection · Data protection, access and sharing · Digital and remote healthcare and research · Research design and adaption · Access to experimental treatments · Regulatory review · Emergency Use Authorisation · Placebo control: social and scientific value · Risk/benefit analysis · Community engagement and participatory processes · Controlled human infection studies · Consent · Sample access and sharing

Global health emergencies such as the COVID-19 pandemic are inevitably both unique and familiar events: some aspects of them can be anticipated and others are unexpected. Outbreaks, epidemics and pandemics have long been part of the human experience and pandemic preparedness is vital. But in any particular emergency the preparations already in place and the assumptions underpinning them will always need to be modified in the light of the emergence of novel pathogen strains and their impacts, changes in human behaviour, and a range of other relevant factors. This suggests an inevitable tension between, on the one hand, the importance of ensuring that health policy and practice in the context of an emergency are informed by evidence gained from previous outbreaks – and by on-going infectious disease science – and, on the other hand, the inevitability that many aspects of a health emergency will be context-specific and unique. Given that knowledge in distinct, complex and evolving emergency situations can only ever be partial and incomplete, research addressing knowledge gaps has a vitally important contribution to make to effective responses to epidemics and pandemics. This highlights the importance both of learning from research conducted during previous pandemics and also of undertaking good-quality research in the current emergency to inform real-time decision-making.

Conducting research in outbreaks, epidemics and pandemics, however, presents an array of complex ethical challenges that need to be identified, analysed and addressed if it is to be successful and capable of commanding well-founded public trust and confidence. Some of these ethical issues will be familiar, but many will be new and/or will arise in combinations not previously encountered. Some of the most important of these ethical considerations arise out of the fact that research needs to be undertaken in circumstances in which societies, health systems, and capacities to conduct research and provide effective oversight of it, are under profound strain.

1.1 Taking a Broader Approach to "Research Ethics"

In this casebook our aim is to encourage an inclusive approach to research ethics which is adequate to address the complexity of conducting research in outbreaks, epidemics and pandemics. In doing so, we build to some extent upon the model set out in the Nuffield Council on Bioethics report on the ethics of research in global health emergencies in 2020, which argued that we need to take a much broader approach to research ethics than has traditionally been the case – both in terms of issues considered to be "ethical", and of the individuals and organizations recognized as having ethical responsibilities ("duty bearers" such as funders, employers and governments). "In brief, ethics is not just about the behaviour of people on the ground, but also the functioning of processes that, however remote they may seem at times to front-line research workers and participants, exert powerful influence on the options actually open to those directly involved in research activities" (Nuffield Council on Bioethics 2020: 4).

Issues that are routinely included within independent ethical review processes, including the need for sensitive study design, well-designed information materials for prospective participants, and meaningful consent processes, are all at the heart of ethically conducted research. Flexible and responsive independent review processes that are sensitive to the time pressures inherent in the emergency are also essential. However, they are not enough. When thinking seriously about the ethical implications of research in emergencies, we need most of all to be alert to fundamental questions relating to the exercise of power and influence, asking ourselves questions such as: "Whose voices (and interests) are steering the choice of priority research topics?"; "Whose voices are determining the way that research is conducted?"; and "Who is going to benefit from this research?". Questions of power and influence arise throughout the trajectory of research, from the scope of an initial funding call to the publication and feedback of findings. Such considerations prompt consideration of diverse aspects of research practice, including:

- The extent to which research is responsive to the multifaceted, differing and inequitable impact of emergencies.
- The ways in which those directly affected by emergencies are engaged in the research process, and the capacity of researchers to respond meaningfully to community input.
- The fairness of research collaborations and capacities of *all* collaborators to influence research priorities, design and conduct, and responsibilities to promote the welfare of front-line research workers.

- Responsibilities arising when seeking to share and access clinical, surveillance, pathogen and research data to support effective pandemic responses.

These considerations also engender ethical responsibilities for a very broad range of duty-bearers; research funders, research institutions, academic publishers, international and intergovernmental organizations, and national governments have policies and practices that profoundly affect how research is conducted.

1.2 Thematic Approach

The cases and chapter themes in this casebook illustrate and provide opportunities for reflection on the importance of a broad, contextually sensitive understanding of research ethics in emergencies. Each chapter includes case studies detailing lived experiences during 2020 and early 2021 to facilitate consideration of the complex ethical issues arising in practice during the COVID-19 pandemic, and their implication for future research in emergencies. The chapter themes are structured to reflect themes arising in the submitted case studies, and do not follow a structure common in research ethics standards, where topics such as research justification, risks and benefits, and consent are addressed sequentially. Chapter 2 addresses the moral complexity of research prioritization in emergencies, including decisions about deprioritising ongoing research. Chapter 3 outlines the importance of conducting rigorous research and appropriately disseminating research findings in non-ideal emergency contexts. Chapter 4 explores the ethical significance of distinguishing between research and non-research activities, such as surveillance, and the challenges that can arise when doing so in rapidly evolving pandemic contexts. Chapter 5 complements this discussion by exploring responsibilities to ensure that research is flexible and responsive to changing policy and evidentiary landscapes. Chapter 6 outlines the issues that can arise when seeking to adapt research review and oversight processes while maintaining substantive participant protections. Chapter 7 explores issues arising when seeking to share and maximise the utility of surveillance, clinical, pathogen and research data to inform pandemic responses. Chapter 8 focuses on responsibilities to ensure that research is appropriately responsive to populations and the inequitable impact of emergencies. Chapter 9 addresses the responsibilities of researchers to ensure that the interests of research participants and communities are appropriately respected when developing and conducting research during public health emergencies.

This first, introductory chapter focuses on the changing, and at times competing, obligations, values and priorities underpinning complex interactions between researchers, regulators and policy-makers during the design, conduct and modification of pandemic research, and early implementation of research findings, against a background of rapidly evolving public health policies and pandemic responses. The four case studies in this chapter illustrate and provide opportunities for reflection on the need for a broad, contextually sensitive understanding of research ethics. When research is prioritized in the pandemic, researchers may need to liaise not just with the typical stakeholders involved in the funding, design, review, conduct and oversight of research, but also to liaise directly with the national authorities who have responsibility for pandemic responses. The cases in this chapter highlight the complex considerations that can arise when research addresses knowledge gaps of national importance during a public health emergency, including adaptions to health-care delivery (Case 1.1), vaccine development (Case 1.2), SARS-CoV-2 infection and pathogenesis (Case 1.3) and COVID-19 epidemiology (Case 1.4). In such cases, direct engagement between researchers and authorities may be necessary for addressing complex ethical considerations, including tensions between prioritizing the conduct and completion of research to address knowledge gaps, accelerating the rollout of novel public health and clinical interventions and approaches, and protecting and promoting the interests of participants and the public.

1.3   The Ethical Importance of Research for Real-Time Policy and for the Benefit of Future Generations

In his book, 'For the common good' Alex London has recently argued that research is not a morally optional activity in a global health emergency (London 2022). Policy-makers, academics, research funders and health systems bear important responsibilities to gather evidence and generate knowledge systematically during emergencies to inform real-time policy-making. They also – importantly – have obligations to future generations. Just as we, to some extent at least, have benefited from research undertaken in previous emergencies, so too will those in the future have a right to expect us to have made systematic attempts at understanding the emergencies we have faced. This suggests that even in situations where it is unlikely that data gathering and analysis are going to inform policy-making during a current emergency, there are nonetheless important obligations to work to inform responses to future emergencies and provide an evidence base for future generations.

This also raises important ethical questions about whether members of the public, patients and so on have moral obligations to participate in such health research. Answers to this question are likely to vary depending on factors such as the risks of participation – how big is the risk and how serious the harm if the risk is realized? It might also depend on the nature of the risk. For example, is it a privacy risk, a risk to physical health, or a risk of some other kind? It seems reasonable to argue that in a situation where the risks were very low, those who are able to contribute to the generation of knowledge and understanding that have the potential to save lives and/or reduce suffering in future generations could have an obligation to do so. A good example might be participation in regular testing programmes or surveys aiming to understand changing levels of infection in the community. There will likely be limits to the risk it would be reasonable to expect anyone to take for others; but what are those limits? What is the nature and scope of the obligations to future generations in the context of an emergency? An important and related question is that of what level of risk in research is it ethically acceptable for researchers to offer to potential participants? Even if it is accepted that people have no obligation to participate in risky research it might perhaps be argued that it is nevertheless ethical to offer them the choice to take part under certain circumstances. Are there limits to the level of acceptable risk? Might this depend to some extent upon the level of background risk to which they are already subject in the context of a pandemic? (Bull et al. 2020).

Quite apart from the obligations of individuals to take part in research, those who develop and implement policy responses to pandemics have an obligation to do so in ways that are capable of generating generalizable knowledge to inform decisions in current and/or future emergencies. During the COVID-19 pandemic, many of the most important public health interventions have not been implemented in ways that have been amenable to either research or rigorous evaluation. Decisions to open or close schools, decisions to mandate face masks, decisions to allow public events have, for example, almost inevitably not been structured in ways capable of providing evidence about their impacts and efficacy. In such cases there are important obligations to act – to set policy – on the basis of such evidence as exists, and to structure policy in ways conducive to generating knowledge – as well as to undertake research as an integral part of public health policy-making in an emergency (Marteau et al. 2022). "Following the science" has been an oft-heard phrase during the COVID-19 pandemic. However, there is also a sense in which policy needs to lead the science by creating the conditions for it and deciding to prioritize research and rigorous evaluation (Massinga Loembé et al. 2020).

The COVID-19 pandemic has also reemphasized that global health emergencies are never only about health. They have wide-ranging impacts, many of which may be as serious as those that affect "health" narrowly defined. Examples might be the impacts we have witnessed on education, on the economy, on employment and on a range of other socially and culturally important activities. This suggests that vitally important research during pandemics should include not only biomedical research but also social science and public health research on, for example, the impact of mask-wearing in schools on social development in young children. This and other possible examples illustrate that research methodologies adequate to making sense of a pandemic will appropriately vary, ranging all the way from classical vaccine studies (Case 1.2), through challenge studies (Case 1.3), to qualitative studies exploring the drivers of vaccine hesitancy (Duong et al. 2022; Momplaisir et al. 2021). In the context of a pandemic, the undertaking of each of these forms of research, both together and in combination, will generate novel ethical questions and require them to be resolved. Ethical issues associated with research prioritisation in a pandemic are addressed in Chap. 2. Examples of ethical issues arising out of the relevance of findings for the subsequent conduct of studies are discussed in Chap. 3 in the context of publication ethics and, in relation to adapting and adaptive trials, in Chap. 5.

1.4 Emergencies as Radically Non-ideal Contexts for Research

Although research is a vital part of pandemic response, in the context of any global health emergency in fact, the reality is that health emergencies are usually radically non-ideal contexts for rigorous research capable of meeting internationally accepted ethical standards. Some of the reasons for this have their origins in the rapidly changing landscape of public health policy and practice in the context of political and other pressures to "do something" or to do with tensions between health and other priorities, such as the economy. Other reasons arise out of the nature of the health emergency itself. As the Nuffield Council on Bioethics states,

> Research in global health emergencies unavoidably takes place in non-ideal circumstances, characterised by disruption, uncertainty, and great health need. This can be compounded by competing claims for legitimacy, time pressures, confusion, and distress. These factors present significant practical challenges to ethical decision-making as practitioners struggle to align their ethical obligations to challenging and often chaotic circumstances. (Nuffield Council on Bioethics 2020, p. 76)

These shifting sands and competing commitments, values and priorities create a context in which rigorous research conducted to high ethical standards may be more difficult to achieve, as discussed in Chaps. 3 and 6. However, it is also these complex, dynamic and interconnected features of an emergency that make it imperative to conduct research to generate the new knowledge required for intelligent public health policy and response. Therefore, research is both important and ethically complicated.

1.5 Relationships Between Research and Practice

The arguments above suggest not only that research in the context of global health emergencies is important but also that such research needs to be part of, i.e. properly integrated within, public health responses to the emergency. It will sometimes, even if not in all cases, need to be based on access to real-time data generated by health authorities and be expected to inform the making of health policy.

Traditionally, research ethics guidelines have taken the view that research and practice should be kept separate for a range of reasons. Perhaps a fundamental concern historically has been to ensure that participants' interests are appropriately protected more broadly (since they are no longer receiving tailored care but instead taking part in an activity designed primarily to gain generalizable knowledge) – given egregious examples where interests have been overridden purportedly for the "greater good" (see Chap. 3). More specifically, many of the arguments for maintaining a clear distinction between research and practice arise out of concerns about the validity of consent if there is the potential for lack of clarity about the distinction between therapeutic and research intentions. People may, for example, mistake research for clinical care. In practice, clear distinctions between research and practice may be blurred for a range of different reasons in different settings (see Chap. 4). In the context of global health emergencies, the appropriateness and achievability of a clear distinction between some forms of public health research and evolving public health activities may be at its least convincing. As Case 1.4 in this chapter demonstrates, distinguishing between activities can be problematized in real time during emergencies, resulting in notable complexities where, for example, the same activity can be designated research in one context or at one time, and as "clinical practice" in others. At the same time, it is important to recognise that such distinctions can have a substantial impact on public confidence during public health emergencies. Chapter 4 includes an interesting example of this in its discussion of the concerns that can arise when there is a lack of clarity about whether a novel vaccine is being offered as a proven public health intervention or as an experimental intervention as part of a clinical trial (Case 4.4 in Chap. 4).

1.6 Modifying Health Policy in the Context of Uncertainty and Open-Endedness

One of the reasons why global health emergencies are likely to constitute radically non-ideal contexts for research is that there is an inevitable mismatch between the timescales of research and those of policy-making. This means that policy decisions of various kinds, e.g. the beginning or ending of lockdowns, emergency use authorization and rollout of vaccines (Case 1.2), or the mandating of mask-wearing on public transport, will often need to be made before definitive evidence is available (even if such a thing is possible). This raises important and urgent questions with a strong value component. What constitutes sufficient evidence? When is the evidence good enough? Tensions such as these between the requirements for acceptable research standards of proof/certainty and those of policy-making in the public interest also, inevitably, require difficult ethical decisions. This is because such decisions will inevitably have implications for morally significant aspects of the lives of those affected by the policy. Policy changes may have implications for equity, or for the liberty or privacy of those affected, or they may have impacts on the well-being of patients and members of the public. People may die or be harmed who would not otherwise have been affected in this way. The making of such policies is ethically difficult where there is good evidence, and even the very best evidence is rarely definitive. Under conditions of uncertainty, ethical complexity is further compounded by morally significant questions relating to the levels of certainty/uncertainty compatible with responsible decision-making. Of course, while there are some situations in which waiting longer might enable the science to progress to a point at which uncertainty was reduced, in many cases – perhaps most – the facts on the ground will change in ways that suggest greater certainty is unlikely in the time available. What does responsible, evidence-based health policy-making look like in such situations?

In other situations, such as in Case 1.1, questions arise not so much in relation to uncertainty but in the context of interventions which research shows to offer "less-than-ideal" solutions. In Case 1.1 the example given is of a technological approach to the remote monitoring of COVID-19 patients at home. What ethical questions are raised by the use of interventions that are considered to be imperfect, or based on uncertain or incomplete data, in the service of public health? Is something better than nothing? If so, when and under what conditions?

1.7 The Impact of Policy Choices on the Ethical Acceptability of Research

There are times when the direction of impact goes the other way, i.e. when changes in policy in response to emerging evidence can raise important questions about the ethics of on-going or proposed research. A good example of this is provided by Case 1.2, in which a placebo-controlled vaccine study is impacted by the authorization of the vaccine under study for public health use under emergency or conditional authorisation. Another example, Case 1.3, is one in which the scientific justification for research judged to be ethically acceptable early in the pandemic, urgent even, required re-evaluation, given the unexpected speed of the development and deployment of new vaccines during the COVID-19 pandemic.

1.8 The Responsibilities of Researchers Who Are Part of the Policy-Making Process

One of the most striking features of the current COVID-19 pandemic has been the high-profile involvement of scientific and social science researchers in the policy-making process. Many governments have claimed in their responses to the pandemics to have been "following the science" and senior scientists have played important and highly visible advisory roles in the making of public policy. In addition to these high-profile roles, many other researchers from a range of disciplines have conducted studies feeding into health policy locally, nationally and internationally. This has the potential to raise interesting and important questions relating to the ethics of scientific practice. Clearly such researchers have obligations to conduct their research to high standards of rigour and to meet the relevant ethical and professional requirements. However, the context of a global health emergency also has the potential to create situations in which difficult questions about the moral responsibilities of scientists arise. This might be because policy appears to contradict the best available evidence, or because evidence is being misinterpreted. It might also be because of a perceived responsibility to counteract the "post truth" aspects of much public, political and media debate. In the context of a public health emergency characterized by great suffering and conflicting beliefs, values and commitments, what are the moral responsibilities of infectious disease scientists, social scientists and other public health researchers?

1.9   Concluding Remarks

The broad approach to research ethics in epidemics and pandemics introduced in this chapter and used in this casebook, aims to capture a comprehensive and context-sensitive range of ethical questions arising in the complex contexts that are inevitably part of global health research on epidemics and pandemics. Our aim in the selection of chapter themes, and the use of cases exclusively drawn from lived experiences, has been to illustrate a range of ethical questions arising during the design and conduct of research, and publication and rollout of research outputs. Against this background, in this chapter we reflect on some of the changing, and at times competing, obligations, values and priorities underpinning complex interactions between researchers, regulators, policy-makers and health authorities when pandemic research is conducted in rapidly evolving research and policy landscapes. We invite reflection on the practical ethical implications of commitments to undertake research during emergencies, including the nature and scope of the relevant responsibilities of a range of stakeholders.

Case 1.1: A Study on the Telemonitoring of COVID-19 Patients at Home

This case study was written by members of the case study author group.

Keywords Boundaries between research, surveillance and rollout; Researcher roles and responsibilities; Resource allocation; Safety and participant protection; Data protection, access and sharing; Digital and remote healthcare and research

As the number of COVID-19 cases surged across the world, many hospitals reached their full capacity to admit and treat COVID-19 patients. In some countries, patients with no symptoms, or mild ones, were asked to quarantine and were assessed remotely by a health-care team, often through phone calls. However, this required a large number of health-care workers to contact patients daily to identify those whose conditions had deteriorated and who required admission to hospital (about 10%). A possible way of improving the monitoring process and reducing health-care workload is to develop automated symptom-monitoring systems. Such systems have been successfully deployed in other areas of medical practice.

Research

Professor E., a digital health expert, received an emergency call from clinical colleagues at a COVID-designated hospital, who asked if she could develop a way of monitoring their patients at home. The hospital had reached its full capacity and could not admit further patients. Those patients who were not admitted were being monitored at home; however, owing to a shortage of staff, they often did not receive calls from the hospital. Professor E. decided to convene a technical team, comprising the top students in her class, to find a solution to this problem. The idea was to develop a remote monitoring system using a chatbot (a computer system designed to simulate interactive conversation with users), which allowed patients to report their symptoms from home. Patients' self-reported data would then be transmitted in real time to a medical dashboard, which could be accessed by the hospital health-care team. The system also had a decision support feature, which prompted patients to call the hospital hotline should they experience new symptoms, or if their condition deteriorated.

As this was a new health application, the team decided to conduct a research project to ensure that it was safe and effective before implementing it on a larger scale. The research team planned to host the system database on a public server to facilitate development, with the intention of transferring it to the hospital once it had been tested. However, the hospital had a data policy which required patient data to be stored on the hospital server. Nevertheless, in view of the nature of the research, both the ethics review committee and the hospital authorities approved the study after the research team assured them that the data would be stored securely and all the developers would sign a non-disclosure agreement. The technical and clinical teams worked tirelessly to develop the e-monitoring system, which was ready for pilot-testing within a week. The clinical team started to use the system with their patients, who were also monitored concurrently by the Hospital Public Health Surveillance Team as part of routine surveillance. While the system reduced the hospital workload significantly, the research team found that half of the patients who had reported that their health had deteriorated failed to call the hospital, and the clinical team still had to reach out to them. Furthermore, because of the overwhelming workload, there were a few near-misses, which were fortunately picked up by the dedicated Hospital Public Health Surveillance Team.

At the end of the study, the research team agreed that, although the system itself was effective and safe in monitoring patients' conditions, implementing it in the real world would require a dedicated team of health-care workers to assess patients remotely to avoid delays in diagnosis and treatment. With the increasing number of patients, the sustainability of the close monitoring of the patients by the health-care team became questionable.

Surveillance

Following this, Professor E. received a call from the state Director of Health, who wanted to use the symptom-monitoring system her team had developed. The Director wanted Professor E.'s team to support the state in monitoring all the COVID-19 patients who were currently being monitored at home. Professor E. had several meetings with the technical and clinical teams at the state Health Office and realized that they did not have the technical expertise or clinical staff to implement the system safely. She also recognized that her technical and clinical teams were too small to cope with the substantial number of COVID-19 patients all over the state.

Professor E. decided to decline the state Director's request but was under tremendous pressure to conform, as she received several calls a day from the state Health Office. She also felt guilty, as she wondered whether offering a less-than-ideal solution was better than no solution, as the patients were not being monitored properly at home anyway. On the other hand, she would feel responsible if the system was implemented without the necessary support, which might lead to a "false sense of security" among both patients and health-care workers, leading to delays in diagnosis and treatment, and possibly even deaths.

Questions
1. What ethical issues should an ethics committee consider when reviewing a digital health system during a pandemic, especially when its potential safety and efficacy are relatively unknown?
2. What types of safeguards for participants might need to be put into place for this type of research and why?
3. Should data security policies be more flexible for research in the context of a global health emergency? Why?
4. What ethical considerations should research teams and public health officials take into account when seeking to find a balance between "perfecting" the monitoring system and rolling it out to address urgent public health needs?

Case 1.2: Trial Unblinding Following Emergency Authorization of Vaccines

This case study was written by members of the case study author group.

Keywords Boundaries between research, surveillance and rollout; Research design and adaption; Access to experimental treatments; Regulatory review; Researcher roles and responsibilities; Emergency Use Authorisation; Vaccines; Placebo control

Vaccines are a crucial tool for preventing and controlling epidemics and pandemics. Consequently, epidemics and pandemics can prompt a flurry of research to develop new vaccines for the pathogen in question. Usually, it takes many years to develop, authorize and deploy a vaccine. However, the COVID-19 pandemic has demonstrated that the vaccine research and development lifecycle – from the first in-human studies to manufacturing, authorization and deployment – can occur so rapidly that vaccines can be used in the same pandemic that prompted the vaccine research in the first place. In part, this may be facilitated through accelerated regulatory mechanisms that permit emergency use authorization (EUA) emergency/conditional authorization (ECA) (i.e. mechanisms that facilitate the public accessibility of investigational vaccines prior to the conclusion of their respective Phase III clinical trials or their licensure) – a process that deviates from the usual approach to vaccine licensing and market authorization (WHO ACT Accelerator Ethics and Governance Working Group 2020).

While laudable, the rapid development, authorization and deployment of novel vaccines in the COVID-19 pandemic raise complex ethical questions and present challenges for ongoing clinical trials for COVID-19 vaccines, due in part to the prospect of continued use of placebo controls in those trials (WHO ACT Accelerator Ethics and Governance Working Group 2020; Singh et al. 2021; Dal-Ré et al. 2021). Randomized, placebo-controlled trials play a key role in research pathways for evaluating the safety and efficacy of novel vaccines (Devereaux and Yusuf 2003). Yet, they raise ethical concerns when participants receiving placebos are deprived of an existing vaccine that is known to be effective against the pathogen in question (CIOMS 2016, see Guideline 5).

When an investigational vaccine receives EUA/ECA, this raises the question of whether participants in the placebo-control arms in ongoing trials for that vaccine are being deprived of an established effective intervention. If so, one might argue that such vaccine trials ought to be unblinded and that participants in placebo-control arms should be offered the vaccine from the trial's experimental arm. On the other hand, one might argue that vaccines granted EUA/ECA do not yet meet the standard of an established effective intervention and that trials with placebo-control arms are still needed after EUA/ECA for purposes such as the following:

- further characterizing and understanding the duration of protection provided by the vaccine
- determining the efficacy of the vaccine in populations not previously included in clinical trials

- evaluating effectiveness for additional clinical endpoints not evaluated in previous clinical trials
- supporting the submission of applications for full market licensure.

This raises a unique challenge for vaccine research and regulation during pandemics, which has implications for national regulatory authorities, vaccine researchers, vaccine manufacturers and those responsible for the ethical review of research.

On 18 December 2020, the US Food and Drug Administration issued emergency use authorization for Moderna's COVID-19 vaccine (US Food and Drug Administration 2020). Nevertheless, a randomized placebo-control trial was set to continue for up to 2 years, to evaluate the clinical efficacy and safety of the vaccine (ModernaTX, Inc. 2020). Americans not enrolled in the trial began to receive the vaccine, and a debate ensued about whether the trial should be unblinded, allowing participants to know whether they were receiving the vaccine or the placebo, so that they could choose to receive the vaccine if they hadn't already done so (Lenzer 2020). Some argued that withholding an effective vaccine from trial participants would be unethical, pointing to the fact that one of the 185 participants who had received the placebo control had died of COVID-19 (Rubin 2021). Others argued that unblinding the trial would compromise its ability to yield reliable scientific data about the vaccine (Cohen 2020). Ultimately, Moderna chose to offer its vaccine to trial participants who had received a placebo (Peres 2021).

Questions
1. If a vaccine receives EUA/ECA for use during a pandemic, what are the competing ethical considerations that need to be addressed for ongoing vaccine research in this context? For example, is it ethical for clinical trials for other vaccines to continue using placebo-control arms? Why?
2. Should vaccines that have been granted EUA/ECA be viewed as an "effective established intervention"? Why?
3. Whose perspectives should matter when making decisions about the continued use of placebo-control arms for vaccine research in this context (e.g., national regulatory authorities, vaccine researchers, vaccine manufacturers, those responsible for the ethical review of research, research participants), and who ought to decide?
4. If the decision is made that it is not ethically required to unblind a trial of a vaccine with ECA, nor to provide the vaccine to the trial participants who received a placebo, is it ethically justifiable to seek to prohibit trial participants from accessing vaccines with ECA for the duration of the research? Why?

References

CIOMS. 2016. *International ethical guidelines for biomedical research involving human subjects*, 4th edn. Geneva: Council for International Organizations of Medical Sciences.

Cohen, J. 2020. Makers of successful COVID-19 vaccines wrestle with options for placebo recipients. *Science*. https://www.sciencemag.org/news/2020/12/makers-successful-covid-19-vaccine-wrestle-options-many-thousands-who-received-placebos.

Dal-Ré, R., W. Orenstein, and A.L. Caplan. 2021. Being fair to participants in placebo-controlled COVID-19 vaccine trials. *Nature Medicine* 27: 938.

Devereaux, P.J., and S. Yusuf. 2003. The evolution of the randomized controlled trial and its role in evidence-based decision making. *Journal of Internal Medicine* 254(2): 105–113.

Lenzer, J. 2020. COVID-19: Should vaccine trials be unblinded? *British Medical Journal* 371: m4956.

ModernaTX, Inc. 2020. *Clinical study protocol: A Phase 3, randomized, stratified, observer-blind, placebo-controlled study to evaluate the efficacy, safety, and immunogenicity of mRNA-1273 SARS-CoV-2 vaccine in adults aged 18 years and older.* https://www.modernatx.com/sites/default/files/mRNA-1273-P301-Protocol.pdf.

Peres J. 2021. Vaccine trial participants who received placebo now hop the line for the real thing for Pfizer, Moderna. *Chicago Tribune*, January 14. https://www.chicagotribune.com/coronavirus/ct-life-covid-vaccine-placebo-pfizer-moderna-janssen-trial-20210114-r6vzmohbs5ed5gstqwux dhtria-story.html.

Rubin, R. 2021. The price of success – How to evaluate COVID-19 vaccines when they're available outside of clinical trials. *Journal of the American Medical Association* 325: 918–921.

Singh, J.A., S. Kochhar, J. Wolff, and the WHO ACT-Accelerator Ethics and Governance Working Group. 2021. Placebo use and unblinding in COVID-19 vaccine trials: Recommendations of a WHO expert working group. *Nature Medicine* 27: 569–570. https://doi.org/10.1038/s41591-021-01299-5.

US Food and Drug Administration. 2020. *Moderna COVID-19 vaccine.* https://www.federalregister.gov/documents/2021/01/19/2021-01022/authorizations-of-emergency-use-of-two-biological-products-during-the-covid-19-pandemic-availability.

WHO Access to COVID-19 Tools (ACT) Accelerator Ethics and Governance Working Group. 2020. *Emergency use designation of COVID-19 candidate vaccines: Ethical considerations for current and future COVID-19 placebo-controlled vaccine trials and trial unblinding.* Policy Brief. Geneva: World Health Organization.

Case 1.3: COVID-19 Controlled Human Infection Studies

This case study was written by members of the case study author group.

Keywords Social and scientific value; Risk/benefit analysis; Safety and participant protection; Ethical review; Community engagement and participatory processes; Controlled human infection studies

Within the context of the COVID-19 pandemic there has been international recognition of the urgent need to develop and distribute safe and effective vaccines globally (United Nations 2020). Vaccine development typically takes 10–20 years, including lengthy trials with human participants for the collection of sufficient evidence about safety and efficacy. In early 2020 calls were consequently made to consider conducting controlled human infection studies (CHIS) (also known as challenge studies) to inform and accelerate vaccine development (Plotkin and Caplan 2020; Eyal et al. 2020) and two COVID-19 CHIS commenced in 2021.

CHIS involve intentionally exposing participants to pathogens in order to study mechanisms of infection and disease and/or the efficacy of experimental vaccines or treatments. CHIS have made important contributions to the treatment and prevention of many infectious diseases (Jamrozik and Selgelid 2021). However, research methods which involve exposing healthy volunteers to a pathogen which can cause infection and disease may seem ethically counter-intuitive – particularly when natural infections with the pathogen can result in severe disease and death. Consequently, in the context of the COVID-19 pandemic, the ethical acceptability of exposing healthy volunteers to SARS-CoV-2 has been the subject of considerable interest and debate in the popular press, social media and academic literature (Callway 2020; Dawson et al. 2020; AVAC and TAG 2020). Two early consensus documents about the ethics of conducting COVID-19 CHIS neither sanctioned nor prohibited such research, but instead highlighted ethical issues requiring careful consideration during the design and review of any such studies (WHO Working Group for Guidance on Human Challenge Studies in COVID-19 2020; Jamrozik et al. 2021; Shah et al. 2020). These drew on broader norms of research ethics and focused on the need for substantial social and scientific value, appropriate risk–benefit profiles, careful site and participant selection, rigorous engagement and consent processes, appropriate compensation, and effective review, oversight and co-ordination.

In early 2020, a consortium of academics, industry collaborators, and the British government (through the Human Challenge Programme of the UK Vaccines Taskforce) began assessing the ethical and practical considerations associated with conducting COVID-19 CHIS. In April 2020, as plans to conduct COVID-19 CHIS progressed, a programme of public consultation and engagement commenced (HIC-Vac 2021). As an initial step, a series of online focus group discussions were held with 57 adults aged between 20 and 40 years. Attendees of the focus group discussions were given an outline of a COVID-19 CHIS protocol and asked about whether they thought the research should be done, what concerns study volunteers might have, how the risks of the research should be explained to potential volunteers, what they felt would constitute acceptable levels of financial compensation for participation, and whether they thought details of proposed studies should be made publicly available (Gbesemete et al. 2020) The results suggested the public was largely positive about the research, despite potential uncertainties about risk levels. During recruitment, the importance of being open and honest both about the risks of the research and about the levels of uncertainty regarding these, was highlighted. Respondents felt that the final decision about the acceptability of proposed COVID-19 CHIS should be made by the scientific community and ethical review committees (Gbesemete et al. 2020).

In December 2020, an application for ethical review of a COVID-19 CHIS, led by researchers from Imperial College, was submitted to the NHS Health Research Authority (Imperial College London 2020). The Authority convened a Specialist Ad Hoc Research Ethics Committee to review the protocol (NHS Health Research Authority 2021). In February 2021, ethical approval was granted to conduct the first COVID-19 CHIS in the world (Department for Business, Energy & Industrial Strategy and Kwarteng 2021). In April 2021 a study led by researchers from the University of Oxford received approval to conduct a second CHIS (University of Oxford 2021). The development and ethical review of both protocols were informed by the World Health Organization's key criteria for the ethical acceptability of COVID-19 CHIS (WHO Working Group for Guidance on Human Challenge Studies in COVID-192020; Jamrozik et al. 2021) and the findings from public consultation and engagement activities. Key considerations arising during the ethical review included the justifications for conducting COVID-19 CHIS, and the appropriate management and minimization of research risks.

Research Justification

At the same time as the COVID-19 CHIS were being designed, vaccine development was progressing exceptionally rapidly. A small number of COVID-19 vaccines began receiving emergency and conditional use authorization before ethical review of the first COVID-19 CHIS took place. As the rollout of these first-generation vaccines began, questions arose about whether COVID-19 CHIS could still be justified. Researchers noted that COVID-19 CHIS still had an important role to play in addressing multiple critical research questions (Rapeport et al. 2021). These included identifying exactly what type of immune responses are needed for effective protection against COVID-19, which would enable accurate testing of whether people's immune responses were likely to protect them from natural infection with COVID-19, and how durable such protection was likely to be. COVID-19 CHIS could also play an integral role in the development of next generation COVID-19 vaccines and of effective treatments, including monoclonal antibody therapies, as new strains of SARS-CoV-2 emerged. This could be especially important where it was necessary to test innovative methods that aimed to provide long-term protection irrespective of viral mutation by stimulating immune responses that were not easily measurable.

Research Risks

International consensus standards for research are clear that risks associated with research should not just be justified solely by the anticipated social and scientific value of research, but should also be reasonable, and appropriately managed and minimized. In the context of the pandemic, the evaluation of the risks of COVID-19 CHIS is complicated by the novelty of the pathogen, and the rapidly changing, and at times contested, evidentiary landscape (Bull et al. 2020). Risk management approaches in COVID-19 CHIS include restricting recruitment to young adult volunteers (aged 18–30) and conducting comprehensive testing and screening to exclude those with underlying conditions and ensure participants are healthy. The CHIS were also designed with cautious dose-escalation to optimize the infection rate if necessary, and started by exposing participants to very small amounts of SARS-CoV-2. Doses would only be incrementally increased if the initial lowest dose of SARS-CoV-2 was insufficient to infect half the participants, and the potential risks of increasing the dose could be appropriately managed and minimized. Any participants who developed COVID-19 symptoms were carefully assessed and received very early antiviral treatment, if required. The studies were conducted within residential quarantine units, where all participants were carefully monitored and discharged once it was confirmed they had not been infected, or if they were no longer infected or at risk of infecting others. Once discharged, participants were followed up for a year (Imperial College London 2020; University of Oxford 2021).

Questions
1. What key ethical issues should be addressed during the design and ethical review of CHIS proposed during an outbreak, epidemic or pandemic?
2. What ethical considerations should be taken into account during the design and conduct of consent processes for COVID-19 CHIS?
3. In studies which seek to address high-priority questions in a pandemic, are higher levels of risk and/or uncertainty about risks more ethically acceptable than in research which does not address such priorities? Why?
4. If controversial research is proposed during an epidemic or pandemic, is it ethically acceptable for it to be developed and reviewed without public consultation and engagement? Why?

References

AVAC, and TAG (Treatment Action Group). 2020. *AVAC and TAG statement on ethical conduct of SARS-CoV-2 vaccine challenge studies.* AIDS Vaccine Advocacy Coalition. https://www.avac. org/blog/avac-and-tag-statement-ethical-conduct-sars-cov-2-vaccine-challenge-studies.

Bull, S., E. Jamrozik, A. Binik, et al. 2020. SARS-CoV-2 challenge studies: Ethics and risk minimisation. *Journal of Medical Ethics*, September 25. https://doi.org/10.1136/medethics-2020-106504.

Callway, E. 2020. Should scientists infect healthy people with the coronavirus to test vaccines? Radical proposal to conduct 'human challenge' studies could dramatically speed up vaccine research. *Nature* 580, March 26. https://doi.org/10.1038/d41586-020-00927-3.

Dawson, L., J. Earl, and J. Livezey. 2020. SARS-CoV-2 human challenge trials: Too risky, too soon. *Journal of Infectious Diseases* 222(3): 514–516. https://doi.org/10.1093/infdis/jiaa314.

Department for Business, Energy & Industrial Strategy, and K. Kwarteng. 2021. *World's first Coronavirus human challenge study receives ethics approval in the UK*, February 17. https://www.gov.uk/government/news/worlds-first-coronavirus-human-challenge-study-receives-ethics-approval-in-the-uk.

Eyal, N., M. Lipsitch, and P.G. Smith. 2020. Human challenge studies to accelerate Coronavirus vaccine licensure. *Journal of Infectious Diseases* 221(11): 1752–1756. https://doi.org/10.1093/infdis/jiaa152.

Gbesemete, D., M. Barker, W.T. Lawrence, et al. 2020. Exploring the acceptability of controlled human infection with SARS-CoV2 – A public consultation. *BMC Medicine* 18. https://doi.org/10.1186/s12916-020-01670-2.

HIC-Vac. 2021. *How and when were the public involved in the study?* https://www.hic-vac.org/public-information/human-infection-studies-coronavirus-covid-19/how-and-when-were-the-public.

Imperial College London. 2020. *COVID-19 Human Challenge Study. Leading the world's first human challenge study for COVID-19*. https://www.imperial.ac.uk/infectious-disease/research/human-challenge/.

Jamrozik, E., and M. Selgelid. 2021. *Human challenge studies in endemic settings ethical and regulatory issues*. Springer briefs in ethics. Cham: Springer.

Jamrozik, E., K. Littler, S. Bull, et al. 2021. Key criteria for the ethical acceptability of COVID-19 human challenge studies: Report of a WHO Working Group. *Vaccine* 39(4): 633–640. https://doi.org/10.1016/j.vaccine.2020.10.075.

NHS Health Research Authority. 2021. *Reviewing the world's first COVID-19 challenge study*, February 25. https://www.hra.nhs.uk/about-us/news-updates/reviewing-worlds-first-covid-19-challenge-study/.

Plotkin, S.A., and A. Caplan. 2020. Extraordinary diseases require extraordinary solutions. *Vaccine* 38(24): 3987–3988. https://doi.org/10.1016/j.vaccine.2020.04.039.

Rapeport, G., E. Smith, A. Gilbert, et al. 2021. SARS-CoV-2 human challenge studies – Establishing the model during an evolving pandemic. *New England Journal of Medicine* 385(11): 961–964.

Shah, S.K., F.G. Miller, T.C. Darton, et al. 2020. Ethics of controlled human infection to study COVID-19. *Science* 368(6493): 832–834. https://doi.org/10.1126/science.abc1076.

United Nations. 2020. "Landmark collaboration" to make COVID-19 testing and treatment available to all. *UN News*, April 24. https://news.un.org/en/story/2020/04/1062512.

University of Oxford. 2021. *Human challenge trial launches to study immune response to COVID-19*, April 19. https://www.ox.ac.uk/news/2021-04-19-human-challenge-trial-launches-study-immune-response-covid-19.

WHO Working Group for Guidance on Human Challenge Studies in COVID-19. 2020. *Key criteria for the ethical acceptability of COVID-19 human challenge studies*. Geneva: World Health Organization.

Case 1.4: Early-Stage Investigations into Infectious Diseases

This case study was written by members of the case study author group.

Keywords Boundaries between research, surveillance and rollout; Data protection, access and sharing; Research design and adaption; Researcher roles and responsibilities; Consent; Sample access and sharing

Early investigations into the transmission of a new infection, such as COVID-19, usually rely initially on surveillance data, which are, ideally, routinely collected by public health authorities as part of standard practice. These data may be shared with researchers when there is an urgent need to develop new knowledge that will benefit the public health response and when the capacity for such sharing exists. Collection of any additional data not considered part of routine surveillance, including biological samples and data from specialized investigations such as genomic sequencing, may require additional protocols. Routine surveillance is likely to change over time to include more or different data as the new pathogen is better understood and tests are developed.

Study X was based on the WHO household transmission investigation protocol and aimed to use enhanced epidemiological surveillance and testing data to assess the transmissibility of SARS-CoV-2 as it began to circulate in the population of a high-income country (WHO 2020). The study identified early confirmed cases of COVID-19 and collected data and swab samples from their household contacts for up to 4 weeks, to provide detailed information about symptoms and transmission risk. Study X was national in its scope, gathering data in partnership with local public health agencies that each had a different way of operating. The experience of using a modified version of the WHO protocol in 2020 highlighted some ethical tensions of relevance to early investigation studies during this and future epidemics.

The implementation of Study X was complicated by the early categorization of study elements as either "public health" or "research". Public health practice involved collecting routine surveillance data; what was "routine" was decided at the level of the local public health agency and differed from place to place. Anything additional was considered research. Following this distinction, data and sample collection were undertaken either as part of the emergency public health response, without explicit ethical approval, or as a research activity requiring standard ethical review and approval. At the beginning of Study X, ethics approval was sought to collect data that were not available through routine public health practice, and was granted. The ethics approval process was smooth; however, governance requirements caused delays.

During the COVID-19 response, public health agencies in the country were challenged by the demands of responding to a nascent infectious disease emergency and many were unable to devote resources to coordination with researchers. In some locations, researchers were able to enrol participants and gather data independently of local public health agencies, by leveraging existing research partnerships, but they continued to face challenges when attempting to access complementary, routinely collected data.

Early investigation studies are, by design, carried out as quickly as possible during times of great uncertainty and change. The types of data and samples collected by local health agencies changed during the period in which the study was conducted, owing to shifting information needs, the availability of new tests and increased capacity. Some of these changes meant that data and samples that were originally classified as "research" elements were now routinely collected. These shifts led to unanticipated consequences. Study X researchers were bound by approved study protocol for the collection of (for example) serological samples, because such samples were additional to routine collection. Later during the study, these samples *were* routinely collected by public health agencies. Study X researchers, however, were not able to obtain serological samples without explicit consent. This meant that samples that were added to routine public health practice after Study X began were not available to researchers.

Early investigations that took place in countries with on-going high-incidence COVID-19 infection generated results about transmission and severity quickly, and new knowledge about preventing infection was widely circulated locally and globally. Consequently, subsequent studies in lower-incidence contexts were taking place while public behaviour, government policy and case management were being altered in response to early findings. This created additional challenges for study eligibility and consistency, with, for example, early cases remaining in their households but later cases being removed to hotel quarantine.

Questions

1. How should research and public health practice be distinguished in studies that rely on routine public health surveillance but that may be enhanced by additional data and samples? What are the implications for ethics approval requirements, particularly given that early investigation studies are time-critical?
2. What responsibilities should public health and/or government agencies have to collaborate with researchers in an epidemic or pandemic?
3. What responsibilities do researchers have to communicate findings quickly and accessibly to public health agencies, including analysis of data derived independently of collaboration with such agencies?
4. In a pandemic, how should the design and conduct of context-specific studies be informed by results from other countries, if at all? Is there a point at which further context-specific research cannot be justified? Why?

Reference

WHO. 2020. *Household transmission investigation protocol for coronavirus disease 2019 (COVID-19)*. Version 2.2, March 23. Geneva: World Health Organization. https://apps.who.int/iris/handle/10665/332040.

References

Bull, S., A. Binik, E. Jamrozik, and M. Parker. 2020. SARS-CoV-2 challenge studies: Risks and ethics (or risk minimisation in context). *Journal of Medical Ethics* 47(12): e79. https://doi.org/10.1136/medethics-2020-106504.

Duong, M.C., H.T. Nguyen, and T. Duong. 2022. Evaluating COVID-19 vaccine hesitancy: A qualitative study from Vietnam. *Diabetes & Metabolic Syndrome* 16(1): 102363.

London, A.J. 2022. *For the common good: Philosophical foundations of research ethics.* New York: Oxford University Press.

Marteau, T.M., M.J. Parker, and W.J. Edmunds. 2022. Science in the time of COVID: Reflections on the Events Research Programme in England. *Nature Communications* 13: 4700. https://doi.org/10.1038/s41467-022-32366-1.

Massinga Loembé, M., A. Tshangela, S.J. Salyer, et al. 2020. COVID-19 in Africa: the spread and response. *Nature Medicine* 26: 999–1003. https://doi.org/10.1038/s41591-020-0961-x.

Momplaisir, F., N. Haynes, H. Nkwihoreze, M. Nelson, R.M. Werner, and J. Jemmott. 2021. Understanding drivers of coronavirus disease 2019 vaccine hesitancy among blacks. *Clinical Infectious Diseases* 73(10): 1784–1789. https://doi.org/10.1093/cid/ciab102.

Nuffield Council on Bioethics. 2020. *Research in global health emergencies: Ethical issues.* London: Nuffield Council on Bioethics. https://www.nuffieldbioethics.org/topics/research-ethics/research-in-global-health-emergencies.

Chapter 2
Setting Research Priorities

Tom Obengo and Jantina de Vries

Abstract Time and resource constraints, combined with competing priorities, mean that research prioritization is a critical ethical consideration in pandemics and emergencies, given the increased need for relevant research findings to address health needs, and the multiple adverse ways that emergencies can impact capacities to conduct research. At international, national and local levels, careful consideration is needed of which research topics should be prioritized and on what grounds. This needs to take into account the ethically significant considerations that should inform prioritization; existing frameworks to guide prioritization decisions; and the consequences associated with prioritizing or de-prioritizing research. The need to prioritize research that is directly responsive to the pandemic may generate debate about which types of research should be prioritised, and within fields of research, which studies should be continued, paused, or re-oriented. In determining which research proposals may have the greatest likelihood of reducing urgent epidemic health burdens, both the nature and distribution of such burdens are key considerations. Epidemics and pandemics typically disproportionately affect the most disadvantaged and vulnerable people in society, highlighting the necessity of inclusive and responsive approaches, which evaluate not just which research approaches have the greatest potential public health benefit, but also the likelihood that they will help address inequities. Key questions also arise when determining if current studies should be de-escalated or stopped, particularly when this may result in highly compromised results. It is also important to consider what obligations arise for research communities (including funders) to pledge to taking the outcomes of research prioritisation processes into account. The case studies in this chapter prompt consideration of how qualitative research into the impacts of isolation should be prioritised, and whether and how research prioritization measures should be responsive to widespread use of traditional medicine and off-label use of medications. The

T. Obengo (✉) · J. de Vries
Department of Medicine, Faculty of Health Sciences, University of Cape Town, Cape Town, South Africa
e-mail: tom.obengo@uct.ac.za

© PAHO and Editors 2024   23
S. Bull et al. (eds.), *Research Ethics in Epidemics and Pandemics: A Casebook*,
Public Health Ethics Analysis 8, https://doi.org/10.1007/978-3-031-41804-4_2

cases also highlight issues that research teams may face as research priorities are re-evaluated in pandemics, including whether and how to redesign proposed research in response to the logistical challenges posed by the pandemic and evolving pandemic research priorities.

Keywords COVID-19 pandemic · Research ethics · Public health emergencies · Research priority setting · Resource allocation · Risk/benefit analysis · Vulnerability and inclusion · Social and scientific value · Researcher roles and responsibilities · Ethical review · Traditional medicine · Research design and adaption · Access to experimental treatments · Non COVID-19 research · Pausing and halting research

2.1 Introduction

Generally speaking, research prioritization is the process of determining which research topics or approaches should be considered a priority for approval, funding and staffing, and it allows for the reasoned allocation of scarce resources (Etti et al. 2021). Priority-setting is an important and common challenge for institutions, governments and funders, as there are always limitations on capacities and resources for research. It is, however, a heightened consideration in epidemics and pandemics, given the increased need for relevant research findings to address pandemic health needs, and the multiple adverse ways that epidemics and emergencies can impact the capacity to conduct research. Whether at the international level or the local level, the idea of research prioritization calls for attention to factors such as which research topics should be prioritized and on what grounds, the ethically significant considerations that should inform prioritization, existing frameworks to guide prioritization decisions, and the risks and burdens that may be associated with prioritizing or de-prioritizing research. A prioritization exercise may also give clarity on whether and when it may be appropriate to prioritize research over the provision of health care during pandemics.

2.2 The Role of Research Prioritization in Epidemics

Epidemics are disruptive of normal life, at the level of both the individual and the community, and research also gets interrupted. Particular challenges that arise in epidemics and which influence priority-setting for research, include resource shortages, especially in acute crises; uncertainty over what type of knowledge will be most useful in addressing the epidemic (clinical, epidemiological, virological, social science or economic data for instance); and the implications of neglecting non-emergency research. In addition, epidemics often give rise to new priorities in medical, scientific or social research, especially if they involve new pathogens or variants. This may cause competition between long-standing and emerging research priorities. For example, as resources get re-allocated to addressing a pandemic, there is a risk that important on-going research and disease-control interventions for illnesses that kill more people than the pandemic does – such as malaria in tropical regions – will be suspended (Weiss et al. 2021). During times where some research institutions suspended all non-COVID-19 research activities, concerns may arise that important research areas that were deprioritized may continue to be perceived as no longer as important as the pandemic abates. The potential neglect of non-pandemic research may cause professional rivalry among researchers and have inequitable impacts on capacities to conduct research and on career progression. Furthermore, in the context of epidemics, an important question arises about whether and when it is justifiable for resources to be spent on research rather than on the delivery of health care. Addressing these challenges is difficult and can be frustrating. It makes priority-setting more complex and demanding of time, energy, critical thinking and resources than it otherwise would be. Research prioritization activities need to balance conflicting priorities and create clarity on these issues.

Although research prioritization during a pandemic may be informed by "the most pressing questions for clinicians and public health professionals" and may involve determining which research is likely going to be of the greatest health benefit to the relevant populations (Etti et al. 2021), other value-based approaches may play equally significant roles. A priority-setting exercise which draws on the views of a broad range of persons involved in addressing the epidemic, can help in the identification and ranking of the main questions surrounding the epidemic, and assist funding organizations and national governments to decide which research should be conducted first and why. In contrast, if research prioritization activities are not undertaken and aligned during pandemics, a number of challenges can arise. In the absence of priorities, scarce resources, including funds, research personnel, facilities, equipment and time, may be used ineffectively in research which is irrelevant or insignificant in the face of the bigger aim of addressing health burdens in pandemics or epidemics. Fragmented approaches may result in a lack of focus on important research areas and poor coverage of key research topics in epidemics.

Epidemics tend to be sudden, unexpected and uncertain – and usually funding is not readily available to address the urgent questions they raise. Furthermore, time is important – understood not only in terms of the time it takes to do good research, but also in terms of the time health workers may have to spend on research, given their clinical care responsibilities. In that context, it is very important to ensure that there is an appropriate decision-making process to help make informed decisions about how scarce resources are to be used for maximum effect. Finally, when a resource-allocation process is administered locally or nationally, it can help identify local research priorities which are likely to promote effective problem-solving in affected communities (Etti et al. 2021). Within countries undergoing pandemics, there are likely to be multiple explicit and implicit priority-setting exercises undertaken at different organizational levels, all of which have some influence. In such contexts, aligning or streamlining research priorities may be important.

2.3 What Considerations Should Inform Research Prioritization?

Setting priorities for research involves considering questions about which kinds of research ought to be supported or conducted in emergency situations. In determining which research is likely going to be of the greatest health benefit to the relevant populations, on the one hand there may be uncertainty about what type of knowledge would be most valuable to address the emergency, and on the other there are assumptions about the relative utility of some types of research over others. The need to prioritize specific research themes related to the epidemic may generate controversies on what is a more significant priority: clinical research versus epidemiology; people's social behaviour versus pathogen mutations; public health versus the economic impact of the epidemic. For instance, there may be a focus on health science research rather than research in the humanities, even though the latter may be equally important for the design of interventions that are widely supported by the people who need to adopt those interventions. Concerns have arisen, for example, that the initial absence of social science research during the West African Ebola outbreak played a role in the design of interventions that were not broadly supported. The absence of local support for the way the interventions had been designed resulted in the continuation of traditional burial practices, and avoidance of clinical care facilities, which contributed to further spread of the epidemic. Social scientists' engagement with the socio-cultural aspects of the epidemic played a role in building a response that contributed to the end of the epidemic, and a key lesson learned was that "understanding social dynamics is essential to designing robust interventions and should be a priority in public health and emergency planning" (Wilkinson et al. 2017). Case 2.1 in this chapter provides an example of a social science study which researchers thought might play an important role in understanding how pandemic burdens are experienced by families of patients with severe COVID-19, and inform the design of protocols for support.

In determining which research has the greatest likelihood of reducing epidemic health burdens in relevant populations, both the nature and distribution of such burdens is a key consideration. Epidemics and pandemics tend to disproportionately affect the most disadvantaged and vulnerable people in society, highlighting the importance of using theoretical approaches that take account of their struggles, not approaches that keep these struggles from view (Nussbaum 2009). Pratt and Hyder (2016) recommend the use of concepts that best reflect moral commitments to perform research focused on reducing health inequalities or systematic disadvantage more broadly, which would lead to the prioritization of research with outcomes that are likely to advance the interests of people who are more disadvantaged and would thus have the greatest potential to increase health equity (Nussbaum 2009; Barsdorf and Millum 2017). Certain types of intervention and forms of research are more likely to benefit those whose overall lives and health situations reflect the worst possible disadvantages, and not just those who will be faced with temporary difficulties at the time of research (Barsdorf and Millum 2017). These considerations demonstrate the importance of evaluating not just which research approaches have the greatest potential public health benefit, but also the likelihood that they will help reduce unjust health disparities and address the most pressing health needs of vulnerable and disadvantaged groups (Pratt et al. 2018).

2.4 Research Prioritization in Practice

When developing robust, inclusive and accountable priority-setting exercises in epidemics, the challenges and tensions outlined above emphasize the importance of carefully thinking about which ethical concepts and values should inform priority-setting, how such decisions should be made, who should be involved, and what interests should be represented. The World Health Organization has developed a three-step process for research prioritization in emergency and disaster-management situations, with each step outlining what actions need to be taken by researchers in such situations (Nasser et al. 2021). The first step involves forming a leadership team, understanding context and collecting necessary data, identifying and engaging with stakeholders, and collecting background information. The second step is to identify research options, decide on what criteria to use to prioritize them, and rank the research options. The third step involves actions after the priority-setting exercise, namely conducting the prioritized research projects, implementing their findings, evaluating the impact of those findings, reporting and publishing the priority-setting exercise, evaluating the process and outcome of the exercise, and feeding the results back to inform future exercises (Nasser et al. 2021). The evaluation and feedback are especially important for informing future prioritization.

Another source of guidance for research priority setting exercises has been developed by the Johns Hopkins COVID-19 Clinical Research Coordinating Committee (CCRCC 2021). These guidelines articulate the overarching principles that should be considered when conducting COVID-19 research. They include the following: scientific and ethical soundness, potential to be informative, minimal risks and burdens, safety and effectiveness, the needs of those affected by COVID-19, room for changes in priorities during the pandemic, and transparency to stakeholders. Case 2.2 demonstrates some of the competing considerations that can arise in such exercises, as an example of research which is prioritized not because the treatment is necessarily anticipated to be effective (given limited evidence and low credibility), but because, problematically, it is being widely prescribed or accessed despite the absence of an appropriate evidence base. In this example a proposed study has the potential to be informative in terms of an anticipated lack of evidence about amantadine's value in treating COVID, but arguably there is no scientific foundation that would typically justify research into the drug's safety and efficacy for such off-label use.

In order to streamline the process and to increase the consistency in priority-setting exercises, Viergever et al. (2010) developed a checklist that incorporates nine common elements of good practice in research priority setting intended to "assist researchers and policymakers in effectively targeting research that has the greatest potential public health benefit". These common elements are context, inclusiveness, information-gathering, planning for implementation, criteria, methods for deciding on priorities, use of a comprehensive approach, transparency and evaluation" (Mador et al. 2016; Viergever et al. 2010). The checklist explains what needs to be clarified in order to establish the context for which priorities are set; it reviews available approaches to health research priority setting; it discusses stakeholder participation and information-gathering; it sets out options for use of criteria and different methods for deciding upon priorities; and it emphasizes the importance of well-planned implementation, evaluation and transparency (Viergever et al. 2010).

The importance of ensuring that priority-setting is appropriately informed and responsive to the context, rather than being a one-size-fits-all or other externally imposed approach, is clear. In the COVID-19 pandemic, the COVID-19 Clinical Research Coalition has been established to, among other things, support the "development of locally identified, context-specific research priorities" (Norton et al. 2021). Case 2.3 is an example of where local knowledge – in this case about the widespread use of traditional medicines to alleviate symptoms of COVID-19 – can inform potential national research priorities.

In low-income settings, such as those found in many parts of Africa, past research experiences in previous epidemics may inform research priority setting. At the Africa Centres for Disease Control, a task force for COVID-19 has worked with experts to identify six key priority areas. These are: epidemiology and surveillance of COVID-19; development of diagnostics; clinical characterization of cases; drug and vaccine clinical trials; investigation of the impact of COVID-19 on the health systems; and social science and policy research ACDC (2021). We can see these research priority areas are broad and may not be easy to fund. The team engaged experts from various research institutes in the six research areas and proposed "a limited number of actionable policy statements". The recommendations from the experts provided further details on each research priority area for ease of understanding and implementation, providing a useful example of how some of the procedural considerations outlined above have been met in practice.

2.5 Challenges in Research Priority Setting

Research-prioritization exercises are not straightforward, and they may be controversial. For instance, in some settings, challenges may emerge which may complicate the research. A first and important challenge relating to research priority setting exercises is that they necessarily risk curtailing academic freedom, especially if some research is de-prioritized (Khumalo et al. 2020). Since epidemics cannot usually be accurately predicted, they take place in contexts where researchers are already engaged in other research activities. New priorities may deprioritize current research and constrain researchers' freedom to engage in a subject they are passionate about and would like to develop to a conclusion.

A second, related, challenge is that research priority setting during an epidemic invariably introduces questions about whether research that is on-going should be de-escalated or stopped. An example is Case 2.4, in which researchers had to postpone research on sexual assault in order to prioritize COVID-19 prevention and infection control. Research can be de-escalated or stopped not just because of priority-setting exercises, but also because the risk–benefit profile has been so altered by the pandemic that it is no longer justifiable to undertake the study. Concerns arise especially on research projects which, if stopped or de-escalated, may lead to highly compromised, or altogether unhelpful, results. A few hypothetical responses may suffice. One response is to perceive research and other academic engagements as dispensable luxuries so that we focus only on epidemic-focused research. Another response is to undertake all the research for which resources are available, but establish a dedicated team looking into nothing but the pandemic. A third response is to de-escalate on-going research to give priority to the epidemic, especially in resource-scarce contexts. This could be combined with a policy that allows for some non-pandemic research to be continued, provided that a request is made and approved by an ethics committee and the relevant institutions. Research that is long-standing – for instance, cohort studies – or where the pausing of research activities constitutes a risk for those involved – for instance, clinical trials that require regular follow-up and where the trial drug cannot simply or easily be replaced with standard clinical care – could then be continued without interruption.

A third challenge relating to research priority setting activities relates to their implementation in practice. If priority-setting activities are not accompanied by a plan for implementation or a broad commitment by the research community (including funders) to consider taking the priorities into account, then the exercise is meaningless. For instance, one study conducted in South Africa concluded that "under one-third of the themes of priority questions developed in the KZN [Kwa-Zulu Natal] research prioritization process were reflected in subsequent research projects. Thus, many areas of health and healthcare considered as priorities remain under-researched" (Khumalo et al. 2020).

The case studies in this chapter highlight how some of the challenges discussed above have manifested in practice. Case 2.1 asks what priority should be placed on conducting qualitative research into the experiences of family members of severely ill and dying COVID-19 patients while isolation measures prohibit in-person visits, and how such research should be conducted. Cases 2.2 and 2.3 demonstrate how research prioritization measures may need to be responsive to widespread use of traditional medicine and off-label use of medications to treat COVID-19, despite the lack of evidence about their efficacy in this context. Cases 2.4 and 2.5 illustrate the issues that research teams may face as research priorities are re-evaluated in pandemics. Case 2.4 prompts reflection on the questions research teams may need to consider when determining whether to completely redesign proposed research in response to the logistical challenges posed by the pandemic and evolving pandemic research priorities. In Case 2.5 questions arise about continuing recruitment into oncology trials in public hospitals, as infection-control measures and anticipated constraints on capacities to provide health care prompt a re-revaluation of research priorities.

Case 2.1: Should Death and Grieving During the Pandemic Be Studied?

This case study was written by members of the case study author group.

Keywords Research priority setting; Risk/benefit analysis; Safety and participant protection; Vulnerability and inclusion; Qualitative research

Among the consequences of the COVID-19 pandemic is that, because of infection-control measures, patients may die without their family and friends being present. Saying goodbye is an important act for patients and their families, and helps to initiate healthy grieving. Rituals, performed either by a priest or other religious figure, or by a friend or family member, can help during this process. This kind of farewell, if performed by an appropriate figure, constitutes an act of respect towards the family and an acknowledgement of the human dignity of the person who has passed away (Consuegra-Fernández and Fernández-Trujillo 2020; Eisma and Tamminga 2020).

Social distancing measures profoundly changed the way we interacted with each other at work, and at home, and the way patients received visits in hospital. During the COVID-19 pandemic, visits were restricted, in order to protect patients and health-care personnel. It was necessary for hospitals to design protocols that allow patients to say goodbye to their loved ones through electronic devices, with the patient supported by a member of the medical team (Wakam et al. 2020). If these painful experiences are not processed in an appropriate and timely manner, they can be traumatic and have negative effects on families and communities in the medium and long term. Disease and death without the presence of family members can be a painful experience for all concerned, including the health-care team.

A group of intensive care unit (ICU) health-care professionals and a mental health team suggested conducting qualitative research to find out about the experiences of people saying goodbye to relatives dying of COVID-19 in hospital, in order to propose measures and protocols for support and care in end-of-life situations. The study would use in-depth interviews to study the experiences and perspectives of relatives of patients who died of COVID-19 in ICUs during the first 6 months of the pandemic. The relatives would be contacted by telephone to ask if they would be willing to participate. The interviews would cover aspects of health care, participants' perceptions of their experience (including being with their loved ones), and the communication and support provided by the medical staff.

The researchers considered that the study posed a minimal risk to participants. In the protocol, when describing the consent process, they did not mention any possible adverse effects except for feelings of sadness and pain related to the memory of the participants' loved ones who had died. The interviews would be carried out by three qualified researchers with training in both qualitative research and conducting in-depth interviews with participants in vulnerable situations. The interviews would be recorded and transcribed, and the audio files would subsequently be destroyed, in order to keep the information confidential and protect the identity of the participants. Participants would be advised that they could opt out of the research at any time.

The study protocol was presented for assessment and approval by a research ethics committee.

Questions
1. Should this type of research be conducted during a pandemic? Why?
2. What are the ethical issues raised by this proposed study?
3. What safeguards should be put in place when seeking to recruit participants who are grieving?
4. Should the research be considered to be minimal risk research? What are the risks of such research and how should they be addressed?

References

Consuegra-Fernández, Marta, and Alejandra Fernández-Trujillo. 2020. La soledad de los pacientes con COVID-19 al final de sus vidas (The loneliness of COVID-19 patients at the end of their lives). *Revista de Bioética y Derecho* 50: 81–98. November 23. http://scielo.isciii.es/scielo.php?script=sci_arttext&pid=S1886-58872020000300006&lng=es&tlng=es.

Eisma M.C., and A. Tamminga. 2020. Grief before and during the COVID-19 pandemic: Multiple group comparisons. *Journal of Pain and Symptom Management* 60(6): e1–e4. https://doi.org/10.1016/j.jpainsymman.2020.10.004.

Wakam G.K., J.R. Montgomery, B.E. Biesterveld, and C.S. Brown. 2020. Not dying alone – Modern compassionate care in the Covid-19 pandemic. *New England Journal of Medicine* 382: e88. https://doi.org/10.1056/nejmp2007781.

Case 2.2: Should Widespread Off-Label Use of Medication Influence Research Prioritisation?

This case study was written by members of the case study author group.

Keywords Research priority setting; Social and scientific value; Resource allocation; Treatment repurposing

In late October 2020, the media in Country X reported that a medication containing amantadine was effective against COVID-19. Amantadine is commonly used to treat neurological disorders, including Parkinson's disease and multiple sclerosis. It also exerts some virustatic activity and was previously used in the prevention and treatment of influenza. A media storm was triggered after one doctor from Country X posted his experience of using it to help treat COVID-19 on the website of a health centre. The doctor's statement indicated that amantadine had helped him and his patients to recover from COVID-19. Moreover, interest in the drug increased further after one politician announced that he had recovered thanks to amantadine.

The initial data about amantadine's potential efficacy with regard to COVID-19 appeared in April 2020. Scientists from Country X observed that some patients who were treated for neurological disorders with this drug and were directly exposed to COVID-19 did not develop severe clinical symptoms of COVID-19. Similar reports were published in Country Y. However, except for some anecdotal and poor-quality evidence from case reports and observational trials with a small number of participants, there were no publicly available data from clinical trials about amantadine's efficacy and safety in treating COVID-19. This raised serious doubts and provoked discussion in the scientific community in Country X. In response, the national health authorities commissioned a review of all available data on the use of amantadine in COVID-19 treatment. The report appeared at the end of November 2020. It was concluded that, because of the limited scientific evidence and its low credibility, there was uncertainty about the efficacy and safety profile of amantadine in COVID-19 treatment. The national health authorities did not recommend that amantadine be used to treat COVID-19 but undertook to monitor emerging data about its use.

Although no strong clinical evidence was reported and amantadine did not appear in the COVID-19 treatment guidelines, many doctors continued prescribing it off-label to treat COVID-19. Consequently, sales of the drug increased significantly. Pharmacists soon began to report problems with amantadine availability for patients with long-term neurological disorders who had been taking it for some time. Compared to September 2020, sales of the drug in October were over three times higher (jumping from 5000 to 17,000 packages a month), and in November they continued to increase. Consequently, in December 2020, the national health authorities decided to introduce limits on sales to ensure on-going access to amantadine for patients with neurological disorders. However, problems with the availability persisted, with problematic consequences. People, especially those who were interested in amantadine as a treatment for COVID-19, began seeking sources of the drug beyond pharmacies. Offers to sell amantadine appeared in social media and online forums.

The national interest in amantadine as a potential treatment for COVID-19 was enormous, leading to a risk of widespread off-label use of amantadine to treat COVID-19 without an appropriate evidence base. As a result, the national health authorities decided to fund two clinical trials to evaluate the safety and efficacy of amantadine for COVID-19, including whether it prevents the development of severe COVID-19 symptoms. The trials were run by two research centres with multiple trial sites in Country X. Starting in March 2021, they sought to recruit about 700 patients in total.

Questions
1. What considerations should be taken into account when setting national priorities for clinical research addressing health needs during a pandemic?
2. Should anecdotal data and increased prescriptions for off-label use of a drug to treat COVID-19 influence which clinical trials should be conducted? Why?
3. Should the limited availability of amantadine for patients with neurological disorders prompt the conduct of research evaluating its efficacy at treating COVID-19? Why?

Case 2.3: Studying the Treatment of COVID-19 Patients with Traditional Medicine

This case study was written by members of the case study author group.

Keywords Research priority setting; Social and scientific value; Researcher roles and responsibilities; Ethical review; Traditional medicine

In the early months of the COVID-19 pandemic there was a lack of evidence about effective biomedical treatments for COVID-19 to inform treatment guidelines. In some countries the absence of effective treatment elicited an unprecedented response from traditional medicine (TM) practitioners and researchers, who sought to evaluate the effect of known traditional or herbal remedies in tackling COVID-19 (Kuntia et al. 2022). In seeking to develop safe and efficacious alternatives to the prevalent biomedical standards of care, many studies focused on prophylaxis or the treatment of mild to moderate COVID-19, and in some instances TM was used as an add-on therapy to complement biomedical treatment. As the efficacy of TM for similar conditions (as described in authoritative TM texts) is not routinely assessed in biomedical preclinical studies, the use of repurposed or novel TM formulations as alternatives or to complement biomedical treatment for COVID-19 requires justification.

Country A has a heritage of traditional systems of medicine for the prevention and treatment of diseases, and well-established formal education systems are in place for these. As previous research had demonstrated that some TM remedies were good immunomodulators (Akram et al. 2018), it was thought that they could play a role in controlling COVID-19 infection. Within Country A initial government advice on treatment schedules for COVID-19 included several potential biomedical treatments, including hydroxychloroquine, ivermectin and Remdesivir, although there was still lack of evidence about whether they could be repurposed to effectively treat COVID-19. These were complemented by guidance about infection-control measures. Government guidance did not, however, address the role of any TM formulations or herbal treatments in treating COVID-19. As health is managed at a state level within Country A, some states permitted the use of TM as a prophylaxis or for treating mild to moderate COVID-19, in the absence of clear evidence about appropriate biomedical standards of care.

An investigator from a biomedical institution submitted a proposal to conduct a randomized, open-label clinical trial using an interdisciplinary (TM and biomedical) approach, as many patients in the region preferred TM formulations. Patients with confirmed mild and moderate COVID-19 would receive a TM formulation in one arm of the study or a government-approved potential biomedical treatment in the other arm. As the TM formulation was a well-known and proven immunomodulator used widely by TM practitioners, the researcher wanted to assess its safety and efficacy in comparison to the government advice about potential biomedical treatments for COVID-19. In country A, specific guidelines have to be followed if biomedical researchers conduct research using TM formulations, and all such research should be conducted in collaboration with TM researchers. The researcher's proposal was approved by the local research ethics committee.

Questions

1. In contexts with well-established traditional systems of medicine, should research into the safety and efficacy of TM remedies in preventing and treating pandemic disease burdens be conducted (particularly in the absence of an evidence base about effective biomedical approaches)? Why?
2. What ethical issues might be associated with incorporating TM into expedited pandemic research pathways?
3. What sort of expertise is required to inform expedited ethical review of interdisciplinary research proposals incorporating TM?
4. Should biomedical investigators conduct research on the safety and efficacy of a TM formulation in a pandemic without the involvement of a TM practitioner or researcher? Why?

References

Akram, M., I.M. Tahir, S.M.A. Shah, Z. Mahmood, A. Altaf, K. Ahmad, N. Munir, M. Daniyal, S. Nasir, and H. Mehboob. 2018. Antiviral potential of medicinal plants against HIV, HSV, influenza, hepatitis, and coxsackievirus: A systematic review. *Phytotherapy Research* 32(5): 811–822. https://doi.org/10.1002/ptr.6024.

Khuntia BK, Sharma V, Wadhwan M Chhabra V, Kidambi B, Rathore S, Agarwal A, Ram A, Qazi S, Ahmad S, Raza K and Sharma G. 2022. Antiviral Potential of Indian Medicinal Plants Against Influenza and SARS-CoV: A Systematic Review. *Natural Product Communications* 17(3). https://doi.org/10.1177/1934578X221086988.

Case 2.4: Research Reprioritization During the COVID-19 Pandemic

This case study was written by members of the case study author group.

Keywords Research priority setting; Social and scientific value; Research design and adaption; Vaccines

Prior to the COVID-19 pandemic, a research team in a sub-Saharan African country was planning to conduct a study on reproductive health with a focus on sexual assault. Sexual assault was an important issue in the community with major consequences. The study team was made up of female physicians with complementary medical specialities. The cross-sectional interventional study aimed to investigate the prevalence, types, risk factors, psychological impact and perpetrators of sexual abuse against adolescents.

The study team prepared to submit their protocol to the national ethics review committee. However, plans to conduct the research were suspended during the first wave of the COVID-19 pandemic in the country, owing to the lockdown and preventive measures put in place to contain and limit the spread of SARS-CoV-2 infection. Not only would data collection not be possible for the study, it was also considered important to focus research efforts on COVID-19, given its status as a public emergency and the associated morbidity and mortality.

The research team decided to conduct a study on COVID-19 infection, prevention and control, given the global scarcity of personal protective equipment (PPE) at the time. Health-care workers were facing one of the worst times of their lives as they encountered COVID-19 patients in emergency rooms and isolation and treatment centres. Studies of the impact of COVID-19 on the well-being of health-care workers, and of their perceptions about COVID-19 were consequently considered to be a priority. In particular it was thought important to investigate the impact of the infection prevention and control measures on health-care workers' well-being, as both the measures and compliance with them could affect their mental health. Study objectives included assessment of doctors' risk perception in relation to COVID-19, the prevention and control measures they practised, and their use of PPE. The study sought to assess health-care workers' well-being and perceptions of local infection prevention and control procedures, using the World Health Organization protocol on the perceptions of health-care workers on infection prevention and control (WHO 2020).

This study was considered very relevant in the context of the pandemic and the researchers developed plans to present the results to policy-makers. The study protocol was approved and the study was conducted. The results of the study were written up and reported at various conferences. Two of the conference abstracts won awards.

Questions

1. When and how should decisions be made about prioritizing COVID-19 research and deprioritizing non-COVID-19 research? What are the ethical implications of such decisions?
2. Should research on health-care workers' well-being in the context of COVID-19 have been prioritized over the planned study involving sexual assault if the sexual assault study could have been completed safely? Why?
3. What can research teams and institutions do to facilitate appropriate research prioritization decisions during a pandemic? What ethical issues should be taken into consideration?

Reference

WHO. 2020. Perceptions of healthcare workers regarding local infection prevention and control procedures for COVID-19: Research protocol. R&D Blueprint COVID-19. Version 1.0. Geneva: World Health Organization. https://www.who.int/docs/default-source/blue-print/perceptions-of-healthcare-workers-protocol-v1–0.pdf?sfvrsn=3f0dd47c_4.

Case 2.5: Challenges with Continuing Cancer Research in a Publicly Funded Hospital

This case study was written by members of the case study author group.

Keywords Research priority setting; Resource allocation; Access to experimental treatments; Non COVID-19 research

A cancer research unit situated within a large, publicly funded hospital in a high-income country ran a wide range of clinical trials of treatments for cancer. Owing to restrictions on the public funding of pharmaceuticals in the country, some cancer treatments that were widely used internationally and of potential benefit to patients were sometimes only available in the context of a clinical trial at the unit. Some of the trials run by the unit involved intravenous treatments and oral agents administered in the same facility in which routine cancer treatments were also provided.

When COVID-19 emerged as a significant threat to populations globally, this country instituted a lockdown prohibiting non-essential movement outside the home. The hospital postponed elective procedures and limited its activities in preparation for an influx of COVID-19 patients, and to reduce the risk to vulnerable patients and staff. Social distancing was instituted and treatment spaces were spread out to increase the distance between patients, reducing the number of patients that wards and units were able to accommodate at any one time.

The treatment facility was able to continue to provide care exclusively to cancer patients during the COVID-19 restrictions, but with reduced capacity. The research unit was faced with a decision about whether to continue enrolling patients into clinical trials involving onsite administration of intravenous treatments and oral agents, or whether to halt recruitment until it became clearer what impact the pandemic would have upon the health system. The team were aware of the need to consider carefully how they allocated resources, including clinical staff and clinic space, as they prepared for the possible effects of the uncontrolled spread of COVID-19. Social distancing limited the number of patients able to receive cancer treatment at any one time, whether for routine care or as part of a clinical trial. Because a single facility undertook clinical and research procedures, conducting research in which participants received onsite oral or intravenous medications could have had knock-on effects for clinical patients, including delayed access to routine chemotherapy. However, halting recruitment to some trials could leave patients who might have benefited from in-trial access to a cancer treatment worse off. In addition, running trials requiring onsite procedures would increase the amount of personal contact between patients, staff and participants, thereby potentially increasing the risk of COVID-19 transmission.

Questions
1. When a pandemic limits available resources, what ethical considerations should influence decisions about whether priority should be given to patients receiving chemotherapy as part of standard clinical care or to participants receiving chemotherapy in a clinical trial?

2. Should patients be told that they might have been eligible for participation in a trial if the pandemic had not impacted clinical resources and created the need for social distancing? Why?
3. Who should decide whether, when or how to reinstitute recruitment to trials involving onsite components such as intravenous treatment? What ethical considerations should they take into account?

References

ACDC. 2021. *Research and development priorities for COVID-19 in Africa*. Addis Ababa: Africa Centres for Disease Control. February.

Barsdorf, N., and J. Millum. 2017. The social value of health research and the worst off. *Bioethics* 31(2): 105–115.

CCRCC. 2021. *Prioritisation process for COVID-19 research involving participant interaction*. Baltimore: Johns Hopkins COVID-19 Clinical Research Coordinating Committee. https://ictr. johnshopkins.edu/covid-research-center/review-committees/ccrcc/.

Etti, M., J. Alger, S.P. Salas, R. Saggers, T. Ramdin, M. Endler, et al. 2021. Global research priorities for COVID-19 in maternal, reproductive and child health: Results of an international survey. *PLoS One* 16(9): e0257516.

Khumalo, G., R. Desai, X. Xaba, M. Moshabela, S. Essack, and E. Lutge. 2020. Prioritising health research in KwaZulu-Natal: Has the research conducted met the research needs? *Health Research Policy and Systems* 18(1).

Mador, R.L., K. Kornas, A. Simard, and V. Haroun. 2016. Using the nine common themes of good practice checklist as a tool for evaluating the research priority setting process of a provincial research and program evaluation program. *Health Research Policy and Systems* 14(1).

Nasser, M., R.F. Viergever, and J. Martin. 2021. *Prioritization of research. WHO guidance on research methods for health emergency and disaster risk management*. Geneva: World Health Organization. https://apps.who.int/iris/bitstream/handle/10665/345591/9789240032286-eng.pdf.

Norton, A., and The other members of the GloPID-R, UKCDR, and COVID-19 Clinical Research Coalition Cross-Working Group on COVID-19 Research in LMICs. 2021. Priorities for COVID-19 research response and preparedness in low-resource settings. *The Lancet* 397(10288): 1866–1868.

Nussbaum, M.C. 2009. Creating capabilities: The human development approach and its implementation. *Hypatia* 24(3): 211–115.

Pratt, B., and A.A. Hyder. 2016. How can health systems research reach the worst-off? A conceptual exploration. *BMC Health Services Research* 16.

Pratt, B., M. Sheehan, N. Barsdorf, and A.A. Hyder. 2018. Exploring the ethics of global health research priority-setting. *BMC Medical Ethics* 19. https://doi.org/10.1186/s12910-018-0333-y.

Viergever, R.F., S. Olifson, A. Ghaffar, and R.F. Terry. 2010. A checklist for health research priority setting: Nine common themes of good practice. *Health Research Policy and Systems* 8. https://doi.org/10.1186/1478-4505-8-36.

Weiss, D.J., A. Bertozzi-Villa, S.F. Rumisha, P. Amratia, R. Arambepola, K.E. Battle, et al. 2021. Indirect effects of the COVID-19 pandemic on malaria intervention coverage, morbidity, and mortality in Africa: A geospatial modelling analysis. *The Lancet Infectious Diseases* 21(1): 59–69.

Wilkinson, A., M. Parker, F. Martineau, and M. Leach. 2017. Engaging 'communities': Anthropological insights from the west African Ebola epidemic. *Philosophical Transactions of the Royal Society B* 372(1721). https://doi.org/10.1098/rstb.2016.0305.

Chapter 3
Research Quality and Dissemination

Sergio Litewka and Sarah Sullivan

Abstract This chapter focuses on issues relating to the rigour and quality of research in pandemic contexts, and the dissemination and publication of research findings. Research is indispensable to inform pandemic responses, including the development of new vaccines and therapeutic possibilities. While these studies are badly needed, public health emergencies present profound ethical challenges for the conduct of research. Key questions arise about whether and to what extent research designs should be adapted to pandemic contexts, including which adaptions may be necessary and which are unjustifiable. Where adaptions are needed, their implications for multiple aspects of research require careful consideration, including the quality of research, participant protections, and potential barriers to recruitment and participation. Challenges may also arise with ensuring that consent to research is informed, and that participants can distinguish between research and the early rollout of interventions in rapidly evolving pandemic contexts. Questions also arise about appropriate responses to studies with smaller sample sizes or other methodological flaws, which are proposed to address urgently pandemic priorities. Pressures to urgently contribute to pandemic evidence bases, including issuing pre-publications and press releases about research results prior to peer review, and dramatically accelerating peer-review processes, raise ethical issues about the dissemination and responses to research findings. The publication of poor quality research, including fraudulent research, contributed to the infodemic in COVID-19, and posed significant challenges for researchers, regulators, and policy makers seeking to develop evidence-informed pandemic responses. Accelerated dissemination of research findings prompts consideration of how to promote research integrity and detect research

---

The original version of the chapter has been revised. A correction to this chapter can be found at https://doi.org/10.1007/978-3-031-41804-4_11

S. Litewka (✉)
Institute for Bioethics and Health Policy, University of Miami Miller School of Medicine, Miami, FL, USA
e-mail: SLitewka@med.miami.edu

S. Sullivan
College of Education and Health Sciences, Touro University California, Vallejo, CA, USA

© PAHO and Editors 2024, corrected publication 2024
S. Bull et al. (eds.), *Research Ethics in Epidemics and Pandemics: A Casebook*, Public Health Ethics Analysis 8, https://doi.org/10.1007/978-3-031-41804-4_3

misconduct, and responsibilities to uphold research quality standards and ensure that publications make constructive contributions in challenging pandemic contexts. The five cases in this chapter promote reflection on citizen-scientists undertaking self-experimentation to develop COVID-19 vaccines outside frameworks for ethical and regulatory review of research; researchers proposing and undertaking research of questionable value and quality with vulnerable populations; and responsibilities of researchers, reviewers, journals and other research during accelerated pre-publication and peer-review processes.

Keywords COVID-19 pandemic · Research ethics · Public health emergencies · Researcher roles and responsibilities · Research publication ethics · Community engagement and participatory processes · Regulatory review · Citizen science · Research misconduct · Vulnerability and inclusion · Safety and participant protection · Social and scientific value · Risk/benefit analysis · Ethical review · Consent · Data protection · Access and sharing

3.1   Introduction

In a pandemic, it is necessary to conduct research and generate evidence rapidly in order to inform effective responses. As the World Health Organization states, "there is an ethical imperative to conduct research during public health emergencies, as some research questions can be adequately investigated only in emergency contexts" (WHO 2020). Such research is indispensable for developing vaccines as well as new therapeutic possibilities. While these studies are badly needed, biomedical research carried out during public health emergencies presents ethical challenges. The five cases in this chapter focus on issues relating to the quality of research, and the dissemination and publication of its findings.

3.2   Research Quality

Research quality standards address the entire research process, including the identification of a research question, the selection of a study approach, data collection and analysis, and the presentation and publication of results (Jacobsen 2016). Research quality considerations are relevant to all aspects of research study design, specifically:

- the match between the research questions and the methods
- participant selection
- outcome measurement

- protection against bias and error (Boaz and Ashby 2003; Shavelson and Towne 2002).

Biomedical research is subject to ethical frameworks based on international consensus standards within the global health community (CIOMS 2016). International guidelines set out ethical principles for the design, aims, revision and follow-up of any research involving human participants, and most countries have integrated some or all these principles into their local regulatory research frameworks as safeguards.

International standards for research ethics seek to protect the interests of human participants and include requirements for research to be reviewed by independent research ethics committees (RECs) (see Chap. 6). Such review includes consideration of the research protocol, including the social and scientific value of the study, and the consent process. Safeguards address the fairness of the recruitment process and the balance of potential benefits and risks for research participants. These safeguards are important to ensure research is ethical and of appropriate quality.

Research quality is a fundamental ethical issue, as the evidence informing responses, practices and policies must be valid and trustworthy. Safeguarding the quality of research is integral to justifying its conduct and ensuring that the anticipated benefits outweigh potential risks and burdens. Research integrity, as a subset of research ethics, urges investigators to possess and steadfastly adhere to "professional standards as outlined by professional organizations, research institutions, and, when relevant, the government and the public" (CIOMS 2016). Research ethics and research integrity are both necessary to ensure research quality during a global health emergency.

A range of pandemic-specific considerations can impact research quality. Concerns related to the value and quality of research during the COVID-19 pandemic have included the following (Glasziou et al. 2020; Lidz and Appelbaum 2002; Luxembourg Agency for Research Integrity n.d.):

- studies being conducted with inadequate scientific background and justification
- studies being conducted with very small sample sizes and limited statistical power
- effective evaluation of the social value of proposed research (pandemic priorities may be a significant multiplier in evaluations of social value)
- an abundance of COVID-19-specific funding resulting in the unnecessary conduct and duplication of studies
- appropriate oversight and conduct of research given gaps in expertise in RECs, on editorial boards, and amongst research teams pivoting to address pandemic priorities

These concerns can all occur in research during pandemics because of the urgency, stress, hype, potential for fame and career progression, and overwhelming need for reliable treatment and prevention of the disease in question.

Additional pandemic-specific study limitations can occur: for example, research quality may be impacted and results biased when specific populations, including vulnerable populations facing increased barriers to accessing health care, are under-

or over-represented in research cohorts (Etowa et al. 2021). Additional challenges arise when research is conducted in hospitals which fail to collect high-quality data because their health workers are overburdened and their systems overwhelmed (Rojek et al. 2020). Furthermore, the urgency created by the pandemic could decrease the scientific rigour required to ensure that research is robust and generates reliable conclusions. This urgency, combined with lapses in research ethics, integrity and study design, have resulted in "a carnage of substandard research" during the pandemic (Bramstedt 2020). To better develop a robust evidence base going forward, researchers, REC members and journal editors, as well as academic and research organizations, should be vigilant about the specific considerations which can impact research quality during public health emergencies (ENRIO 2020).

Suggestions that the urgency generated by a pandemic justifies making changes to research standards (pandemic exceptionalism) must be treated with caution. Should we change research standards during a pandemic because of the magnitude of health burdens and urgent need for evidence to inform responses? Concerns have arisen that in practice, the impact of COVID-19 and a scarcity of qualified peer reviewers may result in a proliferation of studies with small sample sizes and other methodological flaws (London and Kimmelman 2020). Two cases in this chapter (Cases 3.1 and 3.2) highlight issues that arise when research standards are bypassed entirely in the context of a pandemic. When the economic crises and losses of human life are devastating, researchers may be tempted to forge ahead with pandemic priorities without adhering to research governance processes, which can be seen to obstruct progress. Case 3.1 addresses issues related to self-experimentation and citizen science, in a situation where researchers self-administered inoculations with the aim of developing a new COVID-19 vaccine and actively sought to work outside standard research ethics and governance processes. In Case 3.2 the ethics of research on a group of incarcerated subjects using an unproven therapeutic alternative are considered in a situation where the research does not comply with international research ethics standards.

Research ethics guidelines stipulate that researchers can only invite participants to consent to research which meets substantive ethical standards, as discussed further in Chap. 6. If a research study's intrinsic value and scientific validity have not been established, and the research participant selection process is not fair, the risk–benefit ratio is not favourable and independent review has not taken place, the study must not be conducted and the participants' consent is ethically irrelevant (Emanuel et al. 2000). Research ethics guidelines prompt researchers to review whether participants understand that they are taking part in research, and that the intervention being tested may not prevent infection or lessen the symptoms of the disease. When the people enrolled in the study are not aware of the difference between being a participant in a clinical trial and being a patient, a situation known as "therapeutic misconception" arises, in which a trial's participants incorrectly believe that they are receiving routine clinical care (Lidz and Appelbaum 2002). Research quality issues specific to the pandemic and which relate to scientific validity, therapeutic misconception and informed consent are further explored in Cases 3.2 and 3.3.

3.3   Disseminating and Publishing Research

Since the start of the COVID-19 pandemic, there have been an unprecedented number of research publications, leading to a new term: infodemic. The World Health Organization describes an infodemic as a time during a disease outbreak when there is too much information, including false or misleading information, in digital and physical environments. It causes confusion and encourages risk-taking behaviours that can harm health, and also leads to mistrust in health authorities and undermines the public health response (WHO n.d.). Social media networks, the mainstream media and scientific journals have highlighted or published studies that lack scientific validity, which have at times influenced clinical decision-making and behaviour. As a result, many educators now recommend a new pedagogy which will enable students to assess the validity of research better and will develop competent consumers and communicators of science information (Nasr 2021).

Peer-reviewed journals have been under pressure to contribute to the pandemic evidence base and have faced many challenges when they have sought to do so. Publication platforms have sought to make research results available quickly, while also promoting review and research quality, which takes time. The peer-review process is considered a fundamental step for assuring methodological rigour and accurate interpretation of research findings; however, weaknesses in the process have become manifest during the pandemic. The increasing complexity of the scientific and research enterprise, coupled with the multidisciplinary nature of many research collaborations, has meant that the reviewer's role has sometimes proved insufficient for analysing every detail of the scientific quality and legitimacy of a research paper.

The pandemic peer-review process was ineffective in many instances, and the scientific community is now debating which actors have which responsibilities to ensure the rigour of publications. Going forward, how can journal editors guarantee that peer review will ensure methodological rigour during global health emergencies? Publication codes of conduct promote integrity, accuracy and rigour, which assist journal editors to make the tough decisions which will assure the quality of the material they publish (Smith et al. 2020; Committee on Publication Ethics 2011). Alternative and creative methods of peer review are also under consideration (Rojek et al. 2020). Examples of challenges arising in processes related to peer review and publication ethics during the pandemic are highlighted in Case 3.4.

Case 3.5 highlights the impact that inadequate peer review can have on the conduct of research. Two articles, accepted in *The Lancet* and *The New England Journal of Medicine* respectively, show that even highly respected journals sometimes lacked the necessary scrutiny for assuring the soundness of the studies they published. One of the papers, the now infamous Surgisphere study, concerned a multinational registry analysis of the use of hydroxychloroquine for treating COVID-19 patients (Mehra et al. 2020a). Almost simultaneously, the same authors published another paper in the *New England Journal*, about cardiovascular disease, mortality and the effect of angiotensin-receptors blockers on COVID-19 patients (Mehra et al. 2020b). In both cases, the articles withstood the peer reviewers' scrutiny, but when they were published, many scientists wondered how it was possible that hospitals from around the world could so easily and expeditiously share the data of thousands of their COVID-19 patients. There was "skepticism as to the integrity and validity of the dataset, statistical analysis, and conclusions" (Lipworth et al. 2020). The authors could not respond to the journal editors' "expression of concern" and request for access to the raw data, so they retracted both articles. It is not clear how fraudulent or other poor-quality research got through the peer-review quality control process used by some high-impact journals during the pandemic. Ineffective peer review processes are further explored in Cases 3.4 and 3.5 in this chapter and are followed by questions that encourage reflection on the roles of researchers, research ethics committees and journal editors during pandemics.

Disseminating poor-quality research can damage both individuals and entire populations. In 1998, a physician, Andrew Wakefield, and some of his colleagues published a paper in *The Lancet* claiming that the measles, mumps and rubella vaccine was associated with the onset of autism in children. The article proved to be fraudulent, but it took 12 years for it to be retracted (Eggertson 2010), and despite the retraction, it is still cited frequently (Suelzer et al. 2019). Wakefield was subsequently stripped of his medical licence, however, the damage he did remains, as measles and its complications have resurged in unvaccinated communities.

The Retraction Watch database, a blog that tracks retractions from scientific journals, showed that between January 2020 and February 2022, 181 articles were retracted or withdrawn by authors, owing to undisclosed conflicts of interests, concerns about data validity or data analysis errors, misleading conclusions, fake peer reviews or duplicative publications (Retraction Watch Database n.d.). The US Office of Research Integrity gives fabrication, falsification and plagiarism as examples of research misconduct. Other regulatory bodies and international agencies now include many other types of behaviour in their lists of detrimental research practices. These practices fall short of being considered misconduct but affect the integrity, reliability and quality of research. They include inadequate research records, neglectful or exploitative research supervision, misleading statistical analysis and, for institutions, a lack of policies for addressing research misconduct allegations (NASEM 2017). The All-European Academies (ALLEA) issued the revised edition of the European Code of Conduct for Research Integrity in 2017. This code delineates a series of principles for good research practices, such as reliability, honesty, respect and accountability, and also characterizes falsification, fabrication and plagiarism as research misconduct. It also addresses emerging challenges emanating from technological developments, open science, citizen science and social media, among other areas, and adds new areas of unacceptable research practice such as manipulating authorship, unnecessarily expanding the study bibliography and misrepresenting research achievements (ALLEA n.d.). Regulatory bodies in many other countries, including Australia, New Zealand, Canada, and some Asian countries, have adopted similar codes of conduct.

In the Symposium on the COVID-19 Pandemic held by the *Journal of Bioethical Inquiry*, scholars offered their pandemic recommendations, which included establishing independent review panels with oversight over the whole research lifecycle, from the methodological study design to the publication stage, outlining clear data-sharing processes, and increasing funding for research facilities and oversight overall. The authors also recommend stricter penalties for research misconduct, more than just the shame of article retraction, and perhaps penalties for complacent research supervision too (Lipworth et al. 2020). In this way the research governance system can be streamlined and adequately funded and respected to facilitate rapid research, while remaining attentive to scientific quality and integrity.

The unprecedented number of COVID-19-related papers submitted as preprints – articles posted online before formal peer review – has been overwhelming. Scientific manuscripts, before going through peer-review processes, were uploaded at an unprecedented pace to preprint sites and widely shared. While open-access preprints represent a way to increase the knowledge of researchers from all around the world and provide opportunities for sharing efforts, the sites that present them risk becoming platforms for the dissemination of poor-quality research and misinformation, and supporting questionable research practices (Bramstedt 2020). Preprints contributed substantially to the COVID-19 infodemic.

A prominent preprint site for health sciences, MedRxiv, was launched in 2019. Founded by Cold Spring Harbour Laboratory, Yale University, and the *British Medical Journal*, MedRxiv aims "to improve the openness and accessibility of scientific findings, enhance collaboration among researchers, document provenance of ideas, and inform ongoing and planned research through more timely reporting of completed research" (MedRxiv n.d.). A search of articles uploaded between 15 January 2020 and 15 February 2022 to the MedRxiv database, using the term "COVID-19", found 15,383 related articles – many of which had not passed peer review – which exacerbated an already confused evidentiary COVID-19 situation. Issues with preprints can be further debated as readers reflect on Case 3.4, which outlines the need for increased research support and improved infrastructure to inform evidence-based responses during pandemics.

However, with proper precautions and oversight, preprints can be a valuable source of information and provide a timely reference hub for the global scientific community during pandemics, as discussed in Case 5.4 of Chap. 5. For example, a frequently cited and shared March 2020 preprint, from researchers at Imperial College London, estimated the effectiveness of lockdown and social distancing measures and played a significant role in informing policy in the United Kingdom at the beginning of the pandemic (Else 2020). Additionally, preprints are a tool for scrutinizing published data and allow readers to alert the authors to methodological inaccuracies that could have led to incorrect conclusions. In one such case, an article about a clinical trial for a COVID-19 vaccine, Epi-Vac Corona published in the *Russian Journal of Infection and Immunity* reported that in a double-blind placebo-controlled trial, the vaccine had developed 100% immunogenicity against the SARS-CoV-2 virus (Ryzhikov et al. 2021). Several scientists who were not involved in that study subsequently communicated in a preprint that the "true immunogenicity of Epi-Vac is lower than claimed" and that furthermore, "it did not lead to the emergence of neutralizing antibodies in healthy volunteers", while two other preprints mentioned the small cohort size of the study, as well as other inaccuracies that rendered the results dubious (Loseva 2022). These preprints quickly alerted the scientific community to the methodological inaccuracies of research that had already been published.

Noting the time-consuming peer-review process for publication in journals, scientific communities are using alternative methods to share information needed to influence practice rapidly during pandemics. Are preprints a channel that meets the urgent need of the scientific community to communicate results while, at the same time, maintaining the standards of quality and plausibility necessary to ensure scientific integrity? How to best supervise and use preprints to share potentially accurate information in pandemics is one of the important discussion questions related to Case 3.4.

3.4   Conclusion

The trustworthiness of scientific research has been at stake during the COVID-19 pandemic, perhaps as never before. The cases in this chapter highlight issues related to research quality and publication for readers to consider and address going forward. The cases underline the need for transparent research processes, where scientists disclose conflicts of interest, sources of funding and study limitations. Such processes foster the responsible conduct of research (Smith et al. 2020). Now is the time for the scientific community to coordinate its activities and uphold the standards necessary to advance research quality and create an environment where high-quality research and publications make constructive contributions, including during pandemics. This chapter invites readers to reflect on case studies that raise several concerns and questions, without easy answers.

Case 3.1: Self-Experimentation in the Development of COVID-19 Vaccines

This case study was written by members of the case study author group.

Keywords Researcher roles and responsibilities; Community engagement and participatory processes; Regulatory review; Research publication ethics; Vaccines; Citizen science; Researcher safety

Soon after the outbreak of the COVID-19 pandemic, scientists in the Americas, Europe and Asia started conducting experiments on themselves in order to develop a vaccine against COVID-19 (Regalado 2020; Murphy 2020). Self-experimentation is in a grey area – it is not addressed directly in key research ethics regulations, including the Declaration of Helsinki, and its legal status is often unclear (Manríquez Roa and Biller-Andorno 2020; Regalado 2020).

A group of researchers, innovators and citizen science enthusiasts in a country in the Americas embarked on self-experimentation with the aim of developing a COVID-19 vaccine. This initiative involved designing, producing and self-administering progressive generations of nasal inoculations. This group was established as a not-for-profit organization and has been sharing its knowledge through their website. They work under open licences without using patents or asserting their intellectual property rights.

According to the members of the group, the rationale behind their initiative is compassion. On their website, until at least March 2021, they claimed that public health, commercial and regulatory infrastructures had so far failed to provide a vaccine to protect humanity against COVID-19. They also stated that during a pandemic, there is an ethical imperative to deploy emergency vaccines as quickly and widely as possible and not to restrict access to information about them to a privileged circle.

This group of scientists released information about how to produce and self-administer their intranasal inoculation. They made publicly available a "white paper": an in-depth report about their product, which contains terms of use, advice about informed consent, goals, technical features, materials, methods, preparation and instructions on how to administer the potential vaccine, as well as an assessment of the immune response in recipients.

This research has not been approved by a research ethics committee. Moreover, the intranasal inoculation was developed without the authorization of the national authority responsible for regulating the development of vaccines. After the publication of an earlier version of the "white paper" in 2020, a professor from a different country offered to produce the nasal inoculation against COVID-19 in his laboratory and to distribute it to the public for free. However, the self-experimenters claim in their website that they cannot guarantee that their nasal vaccine is safe, and that although preliminary assays have shown positive indications regarding efficacy, this requires ongoing confirmation that will be available in another "white paper". The latest version of their "white paper", released in September 2021, states that no expectation is given concerning efficacy in granting protection against SARS-CoV-2. The group of scientists behind this self-experimentation project claim that hundreds of people have self-administered the product and provide a map with researchers based in more than 20 countries across the world who are interested in collaborating on this vaccine development.

Questions

1. Is self-experimentation ethical in the development of vaccines or therapies during a pandemic? Why or why not?
2. What role should national systems for ethical and regulatory review play with respect to self-experimentation during a pandemic? Is it ethical to involve citizen scientists in such a project? Why?
3. Is it ethical to release a product formula to the public without knowing if it leads to the creation of antibodies against COVID-19 in humans? Why or why not?
4. In the context of the development of COVID-19 vaccines or therapies, are there ways in which self-experimentation should contribute to making science a more inclusive activity? Why?

References

Manríquez Roa, T., and N. Biller-Andorno. 2020. Going first: The ethics of vaccine self-experimentation in Coronavirus times. *Swiss Medical Weekly* 150: w20415. https://doi.org/10.4414/smw.2020.20415.

Murphy, H. 2020. These scientists are giving themselves D.I.Y. Coronavirus vaccines. *New York Times*, September 1, updated 8 Sept 2020. https://www.nytimes.com/2020/09/01/science/covid-19-vaccine-diy.html%20.

Regalado, A. 2020. Some scientists are taking a DIY coronavirus vaccine, and nobody knows if it's legal or if it works. *MIT Technology Review,* July 29. https://www.technologyreview.com/2020/07/29/1005720/george-church-diy-coronavirus-vaccine/.

Case 3.2: Research with Chlorine Dioxide in a Prison During the COVID-19 Pandemic

This case study was written by members of the case study author group.

Keywords Researcher roles and responsibilities; Research misconduct; Vulnerability and inclusion; Safety and participant protection; Social and scientific value; Risk/benefit analysis; Ethical review

During the early stages of the pandemic, when evidence about effective treatments for COVID-19 patients was urgently needed, many politicians, health workers, journalists and other influential leaders across the Americas proposed the administration of substances whose effectiveness for preventing or treating the disease was not supported by credible scientific evidence (US Food and Drug Administration 2020b; Gigova 2020; Forgey 2020; Casado et al. 2021).

Large COVID-19 outbreaks were common in prisons across Latin America between June and September 2020. As a result of overcrowding, poor ventilation, limited access to water, and other unsanitary conditions, and prison inmates and their guards were at high risk of contracting COVID-19 (Blakinger and Hamilton 2020; Associated Press 2020). Newspapers featured stories of inmates protesting and pleading for protection from the rapidly spreading virus (Vivanco and Muñoz 2020). To address this situation in a South American city, regional health authorities met with prison officials to discuss possible interventions. After the meeting, the health authorities announced a new trial to test the effectiveness of chlorine dioxide as a treatment for inmates and prison guards with COVID-19 symptoms. Chlorine dioxide is a bleaching product commonly used for water disinfection. A research team of university biochemists and regional health authorities would lead the investigation.

The team developed and finalized their research protocol. Several other university researchers were invited to join the team, but declined to do so, and provided feedback that they felt that the research protocol was neither clear nor appropriate. The research protocol was not reviewed by a research ethics committee, and university authorities did not respond to a request to officially sponsor the study. Despite this, the study went ahead. Despite a lack of evidence about the therapeutic value of chlorine dioxide, its use was common across the country at the time of the research, especially among the poor and marginalized populations, whose access to high-quality health care was limited. Charismatic politicians, journalists and other leaders around the country were promoting chlorine dioxide as a cure for COVID-19 on the radio, television and social media.

In the study, doses of chlorine dioxide were the main intervention and were provided over a period of 25 days to 30 inmates and prison guards with COVID-19 symptoms. According to local and international news reports, the inmates and guards gave their "informed consent" to participate. The study was designed as a pharmacotherapy follow-up study using the Dader method, a common methodology in pharmacology to create standards of practice. All the research subjects recovered from their COVID-19 symptoms during the study. There were no hospitalizations or new cases diagnosed in the prison during the study period. Twenty-one subjects (70%) left the research study after their symptoms resolved but before the 25 days were completed. The reasons for their withdrawal are not known and many other details about the study were not recorded or made public.

After the study was completed, the research team decided not to publicly share their research results. During personal communication between the author of this case study and the lead investigator, the latter mentioned that it was "possible that there were oversights in designing and conducting this experimental research study but the team had good intentions and are open to any recommendations to develop another study going forward".

Before the study, the United States Federal Drug Administration and the Pan American Health Organization had officially warned against the use of chlorine dioxide to prevent or treat COVID-19, citing significant risks of adverse health effects (US Food and Drug Administration 2020a; PAHO 2020a).

Questions

1. What ethical issues were raised by this research?
2. Which concerns should be addressed by researchers when planning research with vulnerable populations during pandemics?
3. What responsibilities do researchers have to assess the trustworthiness and scientific rigour of information about innovative treatments before conducting research with them during pandemics?
4. How should research ethics committees facilitate the ethical conduct of research during health emergencies and pandemics?

References

Associated Press. 2020. Fear – and the coronavirus – spreads through Latin America's unruly prisons. *Los Angeles Times*, April 28. https://www.latimes.com/world-nation/story/2020-04-28/virus-spreads-fear-through-latin-americas-unruly-prisons.

Blakinger, Keri, and Keegan Hamilton. 2020. *"I begged them to let me die": How federal prisons became Coronavirus death traps*. The Marshall Project. June 18. https://www.themarshallproject.org/2020/06/18/i-begged-them-to-let-me-die-how-federal-prisons-became-coronavirus-death-traps.

Casado, Leticia, Ernesto Londoño, and Adam Rasgon. 2021. As Covid deaths soar in Brazil, Bolsonaro hails an untested nasal spray. *New York Times*, March 31. https://www.nytimes.com/2021/03/06/world/americas/brazil-covid-bolsonaro-nasal-spray.html.

Forgey, Quint. 2020. Trump gets stung from all sides after floating injections of disinfectants. *Politico*, April 24. https://www.politico.com/news/2020/04/24/lysol-maker-warns-against-injecting-disinfectants-trump-coronavirus-theory-206268.

Gigova, Radina. 2020. Lawmakers push toxic disinfectant as Covid-19 treatment in Bolivia, against Health Ministry's warnings. CNN, July 20. https://edition.cnn.com/2020/07/29/americas/bolivia-disinfectant-covid-19-intl/index.html.

Nuffield Council on Bioethics. 2020. *Research in global health emergencies: Ethical issues*. London: Nuffield Council on Bioethics. https://www.nuffieldbioethics.org/publications/research-in-global-health-emergencies.

PAHO. 2020a. *PAHO does not recommend taking products that contain chlorine dioxide, sodium chlorite, sodium hypochlorite, or derivatives.* Pan American Health Organization. 16 July. https://www.paho.org/en/node/72109.

PAHO. 2020b. *Ethics and research 10 key points about research during the pandemic.* Washington: Pan American Health Organization. https://www.paho.org/en/documents/infographic-ethics-research-10-key-points-about-research-during-pandemic.

US Food and Drug Administration. 2020a. *Coronavirus (COVID-19) update: FDA warns seller marketing dangerous chlorine dioxide products that claim to treat or prevent COVID-FDA News Release.* April 8. https://www.fda.gov/news-events/press-announcements/coronavirus-covid-19-update-fda-warns-seller-marketing-dangerous-chlorine-dioxide-products-claim.

US Food and Drug Administration. April 2020b. *Coronavirus (COVID-19) update: Federal judge enters temporary injunction against Genesis II Church of Health and Healing, preventing sale of chlorine dioxide products equivalent to industrial bleach to treat COVID-19. FDA News Release.* https://www.fda.gov/news-events/press-announcements/coronavirus-covid-19-update-federal-judge-enters-temporary-injunction-against-genesis-ii-church.

Vivanco, José Miguel, and César Muñoz. 2020. How to prevent Covid-19 outbreaks in Latin America's prisons. News. May 21. Human Rights Watch. https://www.hrw.org/news/2020/05/21/how-prevent-covid-19-outbreaks-latin-americas-prisons#.

Case 3.3: Evaluating the Role of the BCG Vaccine as a Prophylactic in Elderly Populations

This case study was written by members of the case study author group.

Keywords Vulnerability and inclusion; Ethical review; Consent; Social and scientific value; Vaccine repurposing

Elderly populations with associated comorbidities like diabetes, hypertension and other chronic illnesses are at a higher risk of contracting COVID-19 and have higher rates of mortality if they do contract it (Daoust 2020). The BCG vaccine is known to protect against respiratory tract infections in children and adults and is included in the childhood immunization programme in many countries. A general observation was reported in the early months of the COVID-19 pandemic that in countries where the BCG vaccination is routine, the incidence of COVID-19 infection and mortality was lower than in countries where the BCG vaccination is not being provided (WHO 2020).

Given that BCG vaccination may protect individuals from severe respiratory illnesses, in 2020 an institution in an Asian country decided to start an open-label clinical trial which aimed to collect evidence about the effectiveness of the BCG vaccination in reducing COVID-19-related mortality in elderly populations. The study would recruit 500 volunteers aged between 60 and 95 years residing in a community with a high incidence of COVID-19 cases. (As an infection-control measure, residents within the community were not allowed to travel outside it.) Apart from age, there were no exclusion criteria, although researchers planned to record which participants had also received the BCG vaccination in childhood. To encourage elderly people to enrol in the trial, advertisements about the study would highlight the potential of the BCG vaccine to reduce COVID-19-related mortality in the elderly. An intradermal single dose of the BCG vaccine would be provided and follow-up conducted on a monthly basis for 6 months to assess morbidity and mortality, or until a positive test result for COVID-19 occurred. Researchers would travel to local health centres within the community to conduct the research rather than asking participants to travel to the research facility. All participants would be tested for COVID-19 at recruitment and when follow-up was completed. Self-reported adverse events would be monitored and progress assessed during monthly follow-ups, which would be conducted by telephone during the lockdown period. Symptomatic individuals would also receive additional COVID-19 tests during the follow-up period.

In this setting, local ethics committee secretariats drew on national guidelines to determine whether submitted proposals should receive expedited review by a subset of committee members, or full committee review. In the context of the pandemic, committees then assessed whether full committee reviews should be fast-tracked, with a turnaround time of 24–72 h. The principal investigator requested that this protocol receive expedited review, despite the requirement that research with vulnerable populations should always be reviewed by the full committee, to safeguard participants' interests.

Questions

1. What ethical issues are raised by the design of the proposed study and how should these be addressed?
2. What ethical considerations should inform fast-track review processes for research with vulnerable populations during a pandemic?
3. During fast-track review processes, what role should the ethics committee have in evaluating whether the study design is appropriate to answer the research question and/or making recommendations to improve proposed research?
4. What ethical issues could arise during the consent processes for this project?

References

Daoust, J.F. 2020. Elderly people and responses to COVID-19 in 27 countries. *PLoS ONE* 15(7): e0235590. https://doi.org/10.1371/journal.pone.0235590.
WHO. 2020. *Bacille Calmette-Guérin (BCG) vaccination and COVID-19*. Scientific Brief. April 12. Geneva: World Health Organization.

Case 3.4: Publication, Pre-publication and Retraction of Research: How a Pandemic Magnifies Concerns About Publication Ethics

This case study was written by members of the case study author group.

Keywords Research publication ethics; Research misconduct; Researcher roles and responsibilities; Pre-prints; Retractions

The potentially harmful influence of medical research findings that turn out to be erroneous because of flawed methodology or fraud can be long-lasting and widespread (Wakefield et al. RETRACTED 1998). Although an integral part of the research process, peer review alone is no guarantee of adequate scrutiny, as observed during the COVID-19 pandemic. In July 2020, a paper was published that linked 5G millimetre waves with COVID-19 (Fioranelli et al. RETRACTED 2020). The paper was subsequently retracted as it "showed evidence of substantial manipulation of the peer review" (Biolife SAS 2020). Though the paper was discredited by the scientific community, it likely contributed to the misinformation spreading rapidly online, which made it necessary for the the World Health Organization to issue a statement after 5G phone masts were vandalized (Kaushik 2020). Another example of peer-reviewed research which has now been retracted was the National Institutes of Health-funded paper published on 8 October 2020 claiming that Nephrite-Jade amulets may prevent COVID-19 (Turkle Bility 2020). It was claimed that the paper had undergone a standard review process, with multiple rounds of revisions agreed between the authors and two expert reviewers before finally being accepted (Jarry 2020, Retraction Watch 2020). The paper was heavily criticized on social media upon publication, resulting in its temporary removal on 5 November 2020 and eventual retraction at a later date (Williams 2020).

Even though peer review is imperfect, it has been viewed as providing an additional level of formal scrutiny, which preprints have not undergone. The practice of uploading preprints existed well before COVID-19 but was not standard in the life sciences despite lengthy publication timeframes (Fraser et al. 2020). However, in response to the urgent global need to share scientific findings, the COVID-19 pandemic has prompted an unprecedented increase in preprint publishing, which has enabled access to research months before it would be published in a peer-reviewed journal (Fraser et al. 2020). This, along with the increased retraction rates in 2020, has magnified existing concerns about the impact of preprints on the integrity of biomedical literature and science (Teixeira da Silva et al. 2021). There are also concerns about the spread of misinformation and the resulting threats to public health (Teixeira da Silva et al. 2021). For instance, lay people, eager to find out about treatment for COVID-19, or ways of preventing it, may have easy access to research that has not been thoroughly vetted, and may draw their own inferences from it. In the midst of a global pandemic, the findings of such preprint studies may be widely disseminated by the media, who do not always appreciate the preliminary nature of such findings or convey it to the audiences they engage with.

Nevertheless, the benefits arising from rapid access to some preprints must also be recognized. For example, when a preprint generated by an eminent research group suggested that dexamethasone had the potential to save the lives of critically ill COVID-19 patients (Horby et al. 2020) (see also Case 5.4 in Chap. 5) the World Health Organization and some health-care providers immediately issued guidance based on these findings (WHO 2020; Mahase 2020). On the day the preprint was released, the WHO supported the use of the corticosteroids in appropriate patient groups, presumably because the preprint findings arose from a large, well-designed clinical trial and were consistent with previously known benefits of corticosteroids in reducing inflammation and immune responses. While the research completed peer review a few weeks later and its conclusions remain undisputed, there remains a concern that many preprint studies do not achieve such levels of scientific acceptance if and when independently reviewed by experts (Añazco et al. 2021).

The publication of preprints of scientific research, and the peer-review process itself are two end components of the immense research enterprise. Questions arise about whether they can be expected to carry all the weight for discerning whether research put forth for publication has merit; or whether funding bodies, ethics committees, and other oversight bodies also bear some responsibility, particularly during pandemics.

Questions

1. How should the benefits of publishing preprints during a pandemic be weighed against the potential harm they may bring about?
2. Where there is a need for rapid dissemination of research findings via preprints, what conditions should be in place during a pandemic to ensure they achieve the scientific standards expected for the research to be of public benefit? How should we balance these competing interests?
3. Which stakeholders in research have roles and responsibilities during the conduct of research and dissemination of findings during a pandemic, and what responsibilities do they have?
4. Should journals actively encourage the publication of preprints? Why or why not? How should biomedical researchers view the practice of submitting preprints?

References

Añazco, D., B. Nicolalde, I. Espinosa, J. Camacho, M. Mushtaq, J. Gimenez, and E. Teran. 2021. Publication rate and citation counts for preprints released during the COVID-19 pandemic: The good, the bad and the ugly. *PeerJ* 9:e10927. https://doi.org/10.7717/peerj.10927.

Fioranelli, M., et al. RETRACTED. 2020. 5G technology and induction of coronavirus in skin cells. *Journal of Biological Regulators and Homeostatic Agents* 34(4). https://doi.org/10.23812/20-269-E-4R.

Fraser, N., L. Brierley, G. Dey, J.K. Polka, M. Pálfy, F. Nanni, and J.A. Coates. 2020. Preprinting the COVID-19 pandemic. *bioRxiv*. https://doi.org/10.1101/2020.05.22.111294. Published in *PLOS Biology*. https://doi.org/10.1371/journal.pbio.3000959.

Horby, P., W.S. Lim, J. Emberson, M. Mafham, J. Bell, L. Linsell, N. Staplin, C. Brightling, A. Ustianowski, E. Elmahi, B. Prudon, C. Green, T. Felton, D. Chadwick, K. Rege, C. Fegan,

S.N. Chappell, T. Faust, K. Jaki, A. Jeffery, K. Montgomery, E. Rowan, J.K. Juszczak, R. Baillie, L.C. Haynes, and M.J. Landray. 2020. Effect of dexamethasone in hospitalized patients with COVID-19 – Preliminary report. *medRxiv*. https://doi.org/10.1101/2020.06.22. 20137273. http://medrxiv.org/content/early/2020/06/22/2020.06.22.20137273.abstract. Now published in *New England Journal of Medicine*. https://doi.org/10.1056/NEJMoa2021436.

Jarry, J. 2020. *This paper argues an amulet may protect from COVID. Should it have been published?* November 5. McGill Office for Science and Society. https://www.mcgill.ca/oss/article/covid-19-critical-thinking-pseudoscience/paper-argues-amulet-may-protect-covid-should-it-have-been-published.

Kaushik, M. 2020. Conspiracy theories blame 5G for COVID-19 as studies on health effects remain scant. *Business Today*, July 2. https://www.businesstoday.in/sectors/telecom/conspiracy-theories-blame-5g-for-covid19-as-health-effects-remain-unclear/story/408639.html.

Mahase, E. 2020. Covid-19: Low dose steroid cuts death in ventilated patients by one third, trial finds. *British Medical Journal* 369: m2422. https://doi.org/10.1136/bmj.m2422.

Retraction Watch. 2020. Amulets may prevent COVID-19, says a paper in Elsevier journal. (They don't.) October 29. https://retractionwatch.com/2020/10/29/amulets-may-prevent-covid-19-says-a-paper-in-elsevier-journal-they-dont/.

Teixeira da Silva, J.A., H. Bornemann-Cimenti, and P. Tsigaris. 2021. Optimizing peer review to minimize the risk of retracting COVID-19-related literature. *Medicine, Health Care and Philosophy* 24(1): 21–26. https://doi.org/10.1007/s11019-020-09990-z.

Turkle Bility, M., et al. WITHDRAWN (2020) Can Traditional Chinese Medicine provide insights into controlling the COVID-19 pandemic: Serpentinization-induced lithospheric long-wavelength magnetic anomalies in Proterozoic bedrocks in a weakened geomagnetic field mediate the aberrant transformation of biogenic molecules in COVID-19 via magnetic catalysis. *Science of The Total Environment*, https://doi.org/10.1016/j.scitotenv.2020.142830.

Wakefield, A.J., et al. RETRACTED. 1998. Ileal-lymphoid-nodular hyperplasia, non-specific colitis, and pervasive developmental disorder in children. *The Lancet* 351(9103): 637–641. https://doi.org/10.1016/s0140-6736(97)11096-0.

Williams, S. 2020. Paper proposing COVID-19, magnetism link to be retracted. *The Scientist*, November 4 and 5. https://www.the-scientist.com/news-opinion/paper-proposing-covid-19-magnetism-link-to-be-retracted-68126.

WHO. 2020. WHO Director-General's opening remarks at the media briefing on COVID-19 – 22 June 2020. https://www.who.int/director-general/speeches/detail/who-director-general-s-opening-remarks-at-the-media-briefing-on-covid-19---22-june-2020.

Case 3.5: Retracted Research: Impacts and Outcomes

This case study was written by members of the case study author group.

Keywords Research publication ethics; Research misconduct; Researcher roles and responsibilities; Ethical review; Regulatory review; Risk/benefit analysis; Data protection, access and sharing; Treatment repurposing; Multi-centre research; Retractions

Ethical standards for publication exist for several reasons: to maintain a high quality of academic output, to enable the public to trust research findings, and to ensure that people receive credit for their work. These standards are being challenged in the time of COVID-19 because of the need and pressure to publish the results of related research quickly to provide evidence to inform responses to the pandemic.

In February 2020 a protocol for a multinational study (Study X) was developed to evaluate whether chloroquine (CQ) and hydroxychloroquine (HCQ) were effective in preventing COVID-19. Owing to the use of CQ and HCQ for rheumatological conditions and for malaria, both as prophylaxis and in mass drug administration, there is a large amount of data supporting the safety of long-term administration of these drugs (White et al. 2020). However, no conclusive evidence of benefit from pre-exposure prophylaxis had so far been produced in relation to COVID-19. Similarly, no other chemoprophylactic agents had been proven to be effective. The rationale behind usage was based on *in vitro* antiviral activity of chloroquine on SARS-CoV and it was unclear if an *in vivo* effect with clinical benefit would be observed (Wang et al. 2020). The study aimed to enrol tens of thousands of healthy volunteers from among frontline health-care workers and staff who had close contact with COVID-19 patients. Following ethical and regulatory approval, recruitment commenced for Study X in an Asian country in April, and a European country in May.

On 22 May *The Lancet* published an article by Mehra et al. entitled "Hydroxychloroquine or chloroquine with or without a macrolide for treatment of COVID-19: A multinational registry analysis" (Mehra et al. 2020a). Based on a database known as the Surgical Outcomes Collaborative (developed by Surgisphere Corporation), the article claimed that the use of HCQ or CQ was associated with "decreased in-hospital survival and an increased frequency of ventricular arrhythmias when used for treatment of COVID-19". The impacts of the article's claim about cardio-toxicity were rapid, significant and widespread: they included suspensions of recruitment into HCQ studies, and changes in national recommendations for the clinical use of HCQ (Blamont et al. 2020).

Shortly after the publication of this controversial paper, researchers in Study X played a key role in developing an open letter to Mehra et al. and *The Lancet* setting out some concerns about the published research. Published on 28 May, the letter raised multiple concerns about the statistical analysis, ethics and data integrity in the published article (Watson et al. 2020). On 4 June the Mehra et al. paper was retracted at the request of three of its four authors, following the refusal of Surgisphere to share the dataset to enable independent peer review (Mehra et al. 2020b). *The Lancet* subsequently announced changes to its peer-review process, which sought to reduce the risk of research and publication misconduct (The Editors of The Lancet Group 2020). For future publications, authors must now declare that more than one author has directly accessed and verified the data reported in the manuscript.

In early June, the national regulator at the initial European site for Study X issued a general requirement that trials using HCQ suspend recruitment pending a review. In late June, following substantial discussion and correspondence between the national regulator and researchers leading HCQ studies, approval was given to recommence recruitment into Study X.

Sites in Africa, Asia and Europe expressed interest in joining Study X. Regulators and local research ethics committees reviewing the study protocol at these sites continued to raise questions about associations between HCQ and cardio-toxicity in healthy volunteers, despite the retraction of Mehra et al. Even in settings where CQ and HCQ are routinely used in clinical care, regulatory agencies which initially had relatively few queries about the study became increasingly precautionary over time. Protracted review processes resulted in some sites missing the opportunity to join the study following reduced COVID-19 incidence levels. Concerns about cardio-toxicity also impacted recruitment – of the 200+ health-care workers in the initial European site who expressed an interest in joining the study, just 25 joined once recruitment recommenced.

More broadly, the responses to Mehra et al. have impacted a range of studies seeking to evaluate the prophylactic effect of CQ and HCQ. Some proposed studies have not received approval, others have dropped proposed CQ/HCQ arms, and others have been unable to reach recruitment targets.

Questions

1. How should research publishers balance their responsibilities to ensure that COVID-19 research findings are both rigorous and rapidly disseminated?
2. In the context of unprecedented scientific pre-publication, publication and retraction rates during the COVID-19 pandemic, what ethical obligations do regulators, ethics committees and researchers have to monitor, evaluate and respond to research findings of potential relevance to ongoing studies?
3. What responsibilities do ethics committees and regulators have to consider the potential consequences of declining to approve research? Should a retracted article influence such decisions? Why?
4. What ethical issues should be considered when communicating with study participants about the reasons for, and implications of, pausing and restarting research in response to emerging research findings?

References

Blamont, M., A. Smout, and E. Parodi. 2020. EU governments ban malaria drug for COVID-19, trial paused as safety fears grow. *Reuters*, May 28. https://www.reuters.com/article/health-coronavirus-hydroxychloroquine-fr/eu-governments-ban-malaria-drug-for-covid-19-trial-paused-as-safety-fears-grow-idUSKBN2340A6.

Mehra, M.R., S.S. Desai, F. Ruschitzka, and A.N. Patel RETRACTED. 2020a. Hydroxychloroquine or chloroquine with or without a macrolide for treatment of COVID-19: A multinational registry analysis. *The Lancet.* May 22. https://doi.org/10.1016/S0140-6736(20) 31180-6.

Mehra, M.R., F. Ruschitzka, and A.N. Patel. 2020b. Retraction – Hydroxychloroquine or chloroquine with or without a macrolide for treatment of COVID-19: A multinational registry analysis. *The Lancet* 395(10240): 1820. June 5. https://doi.org/10.1016/S0140-6736(20)31324-6.

The Editors of The Lancet Group. 2020. Learning from a retraction. *The Lancet* 396(10257): 1056. https://doi.org/10.1016/S0140-6736(20)31958-9.

Wang, M., et al. 2020. Remdesivir and chloroquine effectively inhibit the recently emerged novel coronavirus (2019-nCoV) in vitro. *Cell Research* 30(3): 269–271.

Watson, James, et al. 2020. An open letter to Mehra et al. and *The Lancet*. Version 4. *Zenodo*, May 28. https://doi.org/10.5281/zenodo.3862789.

White, N.J., et al. 2020. COVID-19 prevention and treatment: A critical analysis of chloroquine and hydroxychloroquine clinical pharmacology. *PLoS Med* 17(9): e1003252.

References

ALLEA. n.d. The European code of conduct for research integrity. https://allea.org/code-of-conduct/.

Boaz, A., and D. Ashby. 2003. *Fit for purpose? Assessing research quality for evidence-based policy and practice*. London: ESRC UK Centre for Evidence Based Policy and Practice. https://www.kcl.ac.uk/sspp/departments/politicaleconomy/research/cep/.../wp11.

Bramstedt, K.A. 2020. The carnage of substandard research during the COVID-19 pandemic: A call for quality. *Journal of Medical Ethics* 46: 803–807.

CIOMS. 2016. *International ethical guidelines for health-related research involving humans*. 4th ed. Geneva: Council for International Organizations of Medical Sciences.

Committee on Publication Ethics. 2011. Code of conduct and best practice guidelines for journal editors. https://publicationethics.org/files/Code_of_conduct_for_journal_editors_Mar11.pdf

Eggertson, L. 2010. Lancet retracts 12-year-old article linking autism to MMR vaccines. *Canadian Medical Association Journal* 182(4): E199–E200. https://doi.org/10.1503/cmaj.109-3179.

Else, H. 2020. How a torrent of COVID science changed research publishing – In seven charts. *Nature*. https://www.nature.com/articles/d41586-020-03564-y.

Emanuel, E.J., D. Wendler, and C. Grady. 2000. What makes clinical research ethical? *Journal of the American Medical Association* 283(20): 2701–2711. https://doi.org/10.1001/jama.283.20.2701.

ENRIO. 2020. ENRIO statement: Research integrity even more important for research during a pandemic. European Network of Research Integrity Offices. http://www.enrio.eu/enrio-statement-research-integrity-even-more-important-for-research-during-a-pandemic/.

Etowa, J., et al. 2021. Difficulties accessing health care services during the COVID-19 pandemic in Canada: Examining the intersectionality between immigrant status and visible minority status. *International Journal for Equity in Health* 20. https://doi.org/10.1186/s12939-021-01593-1.

Glasziou, P.P., S. Sanders, and T. Hoffmann. 2020. Waste in Covid-19 research. *British Medical Journal* 369. https://www.bmj.com/content/369/bmj.m1847.

Jacobsen, K. 2016. *Introduction to health research methods*. Jones & Bartlett Learning.

Lidz, C.W., and P.S. Appelbaum. 2002. The therapeutic misconception: Problems and solutions. *Medical Care* 40(9 Suppl): V55–V63. https://doi.org/10.1097/01.MLR.0000023956.25813.18.

Lipworth, W., M. Gentgall, I. Kerridge, et al. 2020. Science at warp speed: Medical research, publication, and translation during the COVID-19 pandemic. *Journal of Bioethical Inquiry*. https://link.springer.com/article/10.1007/s11673-020-10013-y.

London, A.J., and J. Kimmelman. 2020. Against pandemic research exceptionalism. *Science* 368(6490): 476–477. https://doi.org/10.1126/science.abc1731.

Loseva, P. 2022. Data and distrust hamper Russia's vaccination programme. *British Medical Journal* 376. https://doi.org/10.1136/bmj.o321.

Luxembourg Agency for Research Integrity. n.d. Ethical & robust research during a pandemic: HOW? Webinar. https://www.youtube.com/watch?v=GUHcwPqpp2s.

MedRxiv. n.d. About MedRxiv. https://www.medrxiv.org/content/about-medrxiv.

Mehra, M.R., S.S. Desai, F. Ruschitzka, and A.N. Patel. 2020a. RETRACTED: Hydroxychloroquine or chloroquine with or without a macrolide for treatment of COVID-19: A multinational registry analysis. *Lancet*. Advance online publication. https://doi.org/10.1016/S0140-6736(20)31180-6. (Retraction published *Lancet* 5 June 2020).

Mehra, M., S. Desai, S. Kuy, T. Henry, and A.N. Patel. 2020b. *New England Journal of Medicine* 382: e102. https://doi.org/10.1056/NEJMoa2007621.

Nasr, N. 2021. Overcoming the discourse of science mistrust: How science education can be used to develop competent consumers and communicators of science information. *Cultural Studies of Science Education* 16: 345–356. https://doi.org/10.1007/s11422-021-10064-6.

National Academies of Sciences, Engineering, and Medicine; Policy and Global Affairs; Committee on Science, Engineering, Medicine, and Public Policy; Committee on Responsible Science. 2017. *Fostering integrity in research*. Washington: National Academies Press. https://www.ncbi.nlm.nih.gov/books/NBK475954/.

Retraction Watch Database. n.d.. http://retractiondatabase.org/.

Rojek, A.M., G.E. Martin, and P.W. Horby. 2020. Compassionate drug (mis)use during pandemics: Lessons for COVID-19 from 2009. *BMC Medicine* 18. https://doi.org/10.1186/s12916-020-01732-5.

Ryzhikov, A.B., E.A. Ryzhikov, M.P. Bogryantseva, S.V. Usova, E.D. Danilenko, E.A. Nechaeva, O.V. Pyankov, O.G. Pyankova, A.S. Gudymo, S.A. Bodnev, G.S. Onkhonova, E.S. Sleptsova, V.I. Kuzubov, N.N. Ryndyuk, Z.I. Ginko, V.N. Petrov, A.A. Moiseeva, P.Yu. Torzhkova, S.A. Pyankov, T.V. Tregubchak, D.V. Antonec, E.V. Gavrilova, and R.A. Maksyutov. 2021. A single blind, placebo-controlled randomized study of the safety, reactogenicity and immunogenicity of the 'EpiVacCorona' vaccine for the prevention of COVID-19, in volunteers aged 18–60 years (phase I–II). *Russian Journal of Infection and Immunity* 11(2): 283–296. https://doi.org/10.15789/2220-7619-ASB-1699.

Shavelson, R.J., and L. Towne, eds. 2002. *Scientific research in education*. Washington: National Research Council, National Academy Press.

Smith, M.J., R.E.G. Upshur, and E.J. Emanuel. 2020. Publication ethics during public health emergencies such as the COVID-19 pandemic. *American Journal of Public Health* 110(7): e1–e2. https://doi.org/10.2105/AJPH.2020.305686.

Suelzer, E.M., J. Deal, K.L. Hanus, B. Ruggeri, R. Sieracki, and E. Witkowski. 2019. Assessment of citations of the retracted article by Wakefield et al with fraudulent claims of an association between vaccination and autism. *Journal of the American Medical Association Network Open* 2(11): e1915552. https://doi.org/10.1001/jamanetworkopen.2019.15552.

WHO. 2020. *Ethical standards for research during public health emergencies: Distilling existing guidance to support COVID-19 R&D*. Policy brief. Geneva: World Health Organization. https://www.who.int/publications/i/item/WHO-RFH-20.1.

WHO. n.d. *Infodemic*. https://www.who.int/health-topics/infodemic#tab=tab_1.

Chapter 4
Boundaries Between Research, Surveillance and Monitored Emergency Use

Teck Chuan Voo and Ignacio Mastroleo

Abstract Responses to outbreaks, epidemics and pandemics involves a heterogeneous set of activities that aim to address threats to public health. In addition to research, non-research activities, such as prevention and control interventions, and surveillance, are conducted. The boundaries between research and non-research responses can rapidly blur during a public health emergency such as the COVID-19 pandemic. There may be common elements between these types of activities, and they may draw on the same resources and infrastructure. Non-research activities, such as surveillance and emergency non-research use of unproven interventions, and research activities must all be undertaken in an ethical manner as components of emergency response. However, care is needed to distinguish between non-research public health activities and research, because research often has considerations and requirements for its ethical conduct which are distinct from non-research public health activities. Research aims to produce generalizable knowledge, and mechanisms such as participant consent and independent ethics review aim to ensure that the rights and interests of research participants are respected. Ensuring that research and non-research activities are appropriately distinguished can additionally promote proper coordination of such activities, and increase trust and social accountability in pandemic responses. Consequently, it is important to distinguish between these different activities on the basis of their primary aim, and to consider whether their implementation is justifiable, based on their aims and the relevant ethical framework for each type of activity, and how they are coordinated as part of the larger collective activity of emergency response and management. Complex questions arise about how the different stakeholders involved in decision-making should make valid and justifiable decisions about whether the response activity is research or non-research. The cases in this chapter

T. C. Voo (✉)
National University of Singapore, Centre for Biomedical Ethics, Singapore, Singapore
e-mail: medvtc@nus.edu.sg

I. Mastroleo
National Scientific and Technical Research Council (CONICET), Buenos Aires, Argentina

© PAHO and Editors 2024
S. Bull et al. (eds.), *Research Ethics in Epidemics and Pandemics: A Casebook*,
Public Health Ethics Analysis 8, https://doi.org/10.1007/978-3-031-41804-4_4

invite consideration about how such decisions should be made, and their implications, in the context of applications to conduct retrospective research into the outcomes of emergency uses of unproven interventions outside clinical trials, and of characterising antibody-testing initiatives and systematic data collection activities as surveillance or research.

Keywords COVID-19 pandemic · Research ethics · Public health emergencies · Boundaries between research · Surveillance and clinical care · Data protection · Access and sharing · Consent · Emergency use authorisation · Researcher roles and responsibilities · Return of results · Ethical review · Safety and participant protection · Consent · Privacy and confidentiality · Risk/benefit analysis · Vulnerability and inclusion

4.1 Introduction

A public health response to an outbreak, epidemic or pandemic (OEP) involves a heterogeneous set of activities that aim to address the threat to public health posed by an infectious pathogen. These activities include non-research activities, such as prevention and control interventions and surveillance, and research activities. Prevention and control interventions can be based on the information and findings produced by surveillance and research, which are forms of data collection activities for different purposes.

Non-research public health activities – commonly called "public health practice" – are routinely implemented in emergency and non-emergency situations. Care is always needed to determine the boundaries between non-research public health activities and research (see Table 4.1). This is because research, including its subset of public health research, has particular considerations and requirements for its ethical conduct which are distinct from non-research public health activities (Otto

Table 4.1 Definitions

Public health practice	"[T]he practice of public health (roughly) consists of collective interventions that aim to promote and protect the health of the public" (Verweij and Dawson 2007, p. 22)
Public health surveillance	"[T]he continuous, systematic collection, analysis and interpretation of health-related data needed for the planning, implementation, and evaluation of public health practice" (WHO 2017, p. 14)
Research	Systematic investigation, including development, testing and evaluation, "designed to develop or contribute to generalizable [scientific] knowledge" (National Commission for the Protection of Human Subjects of Biomedical and Behavioral Research 1978, cited in Beauchamp and Saghai 2012, p. 52)
Public health research	"Investigations of interventions in, or studies of, populations, that are anticipated to have an effect on health or on health inequity at population level" (Lockwood and Walters 2018, p. 673)

et al. 2014). Non-research public health activities aim at protecting and promoting the health of a given population and are guided by considerations such as health maximization, mitigation of health inequities, and proportionality, i.e. interventions that infringe on individual rights and interests should be proportional to the relevant threat and risks, and expected health benefits. Research aims to produce generalizable knowledge, and mechanisms such as participant consent and independent ethics review aim to ensure that the rights and interests of research participants take precedence over the production of generalizable knowledge.

As the case studies in this chapter show, the boundaries between research and non-research response can easily and quickly blur during a public health emergency such as the COVID-19 pandemic. One reason for this is the existing or potential common elements between these activities. For example, surveillance and research may apply similar methodologies of systematic data collection, analysis, and dissemination, and can raise similar ethical issues, such as the exposure of individuals to privacy and data protection risks.

Another reason for the blurring of boundaries is that resources and infrastructures set up for research can be co-opted to support non-research public health activities, and vice-versa. For example, existing databases and biorepositories for research on other infectious diseases were used for surveillance purposes during the COVID-19 pandemic (Doerr and Wagner 2020). Research ethics committees (RECs) or institutional review boards – independent ethical oversight mechanisms for research – may also be used to review the emergency provision of unproven medical interventions outside of clinical trials, in accordance with the MEURI (Monitored Emergency Use of Unregistered and Investigational Interventions) ethical framework advanced by the World Health Organization (WHO) (PAHO 2020; WHO 2018) (See Table 4.2).

Table 4.2 The MEURI framework

I. Justification
No proven effective treatment exists.
It is not possible to initiate clinical studies immediately.
Data providing preliminary support of the intervention's efficacy and safety are available, at least from laboratory or animal studies, and use of the intervention outside clinical trials has been suggested by an appropriately qualified scientific advisory committee on the basis of a favourable risk–benefit analysis.
II. Ethical and regulatory oversight
The relevant country authorities, as well as an appropriately qualified ethics committee, have approved such use.
Adequate resources are available to ensure that risks can be minimized,
III. Consent process
The patient's informed consent is obtained.
IV. Contribution to the generation of evidence
The emergency use of the intervention is monitored, and the results are documented and shared in a timely manner with the wider medical and scientific community.

Source: PAHO (2020)

This chapter considers a central theme that runs through the four case studies: the ethical importance of clear boundaries between research and non-research activities – with focus on surveillance and non-research use of unproven interventions – conducted in response to an OEP. This raises two issues. The first is the challenge of identifying whether an activity constitutes research or not (see Table 4.2). This is not a new issue (Barrett et al. 2016; Otto et al. 2014; Taylor 2019; WHO 2015), and has been clearly recognized in previous OEP emergencies, such as the Zika outbreak in Latin America (PAHO 2016), as well as in non-OEP situations (Beauchamp and Saghai 2012; Mastroleo and Holzer 2020). The second issue is the need to coordinate non-research activities with research activities for an effective and ethical OEP response. This invites the question of whether a non-research activity ought to be conducted as research (and thus be justified or guided by a different set of ethical considerations), and vice versa. This chapter will examine these issues and provide key considerations for addressing them as they arise in the contexts of the case studies.

As an important preliminary consideration, it should be noted that there are differing positions on the boundary between research and non-research activities as part of an emergency response to an OEP. One position is that the boundary is not always clear and not all activities within an OEP response can be classified neatly as either research or non-research, for example the monitored use of unproven interventions outside of clinical trials, which has elements of both practice and research. So for activities that lie in the so-called 'fuzzy middle', the focus should be on identifying and addressing the ethical issues, rather than trying to place them into one category or the other. The downside of such a position is that the responsible agent of the activity and the relevant authority would likely find it challenging to determine what ethical standards or safeguards to uphold, or the ethical purpose of procedures like consent and ethics review. A further potential effect is that policymakers may set an arbitrary set of ethical or regulatory rules for some activity without adequate justification. Alternatively, they may let the responsible agents figure out what to do by themselves in an uncoordinated manner, which risks undermining an effective and ethical response to a public health emergency.

The second position on the boundary is that an activity that has both research and practice components and multiple aims can be categorized as either research or not according to its *primary aim*, which gives the activity its ethical character and orients its requirements. Insofar as a reasonable national response to an OEP emergency depends on appropriate ethical regulation of activities and their transitions, which in turn depends on how they are classified, such an activity ought to be classified as either a research or non-research activity rather than be left in the fuzzy middle. This also applies to monitored emergency use of unproven interventions outside of a clinical trial, which one of us have argued should be classified as an emergency care practice according to its primary aim (as discussed below). "Fuzziness" or "grey areas" may be cases where individuals and institutions are confused about (1) the primary aim of an activity, (2) its place in the overall response to an OEP, (3) how to design that activity according to its primary aim, or (4) if the design of an activity is appropriate given the classification of the activity and its multiple aims (which ought to be subordinated to the primary aim). These issues could be resolved by an appropriate mechanism of review by a national bioethics committee, or expert consensus backed by a competent national authority in charge of the emergency response. In the case of international public health emergencies, such a mechanism should draw on guidance from international documents and expert committees convened by international health authorities (e.g. WHO, Pan American Health Organization (PAHO)) where national states participate as members. The discussion in this chapter is based on the second position, which we support.

4.2   The Research–Practice Distinction

Research involving human subjects, including during public health emergencies, has to uphold appropriate scientific and ethical standards (London and Kimmelman 2020; WHO 2020a) by observing requirements such as scientific validity and value, social value, independent ethics review, privacy and confidentiality, the right of subjects to withdraw from the research, and informed consent (Emanuel et al. 2008).

Case 4.1 raises the issue of whether consent should be waived with respect to a retrospective study on the clinical outcomes of emergency use of convalescent plasma for treating COVID-19. REC approval of consent waiver, which typically applies to secondary research using individually identifiable information, depends on the satisfaction of some or all of the following considerations: the study meets some threshold of serving the public good or public interest; it is necessary to use individually identifiable information; it is impracticable to obtain informed consent; the study presents no more than a minimal risk to the participants; and the waiver will not adversely affect their rights and welfare (Schaefer et al. 2020). Adherence to research ethics requirements and standards ensures that use of the information and bodily materials of individual persons to achieve research goals will not take precedence over their rights and interests, which is important for safeguarding public trust in research as a scientific endeavour.

In comparison, although individual rights and interests should be respected, it is sometimes necessary, and legitimate, for non-research public health activities to override these rights and interests in order to fulfil the public health mandate of protecting and promoting the collective health of a community. Consider Case 4.2 which discusses the ethical justification for decisions on the return of individual antibody test results by a public health surveillance testing initiative and two seroprevalence research initiatives. Broadly, assuming that the tests used are validated and highly accurate, public health practitioners conducting surveillance testing are justified in returning individual test results for public health reasons (e.g. as the basis for restriction of movement to prevent transmission), even if there is no prior consent. In comparison, researchers' ethical obligation to disclose individual test results in the context of seroprevalence research (as well as many other types of research) would depend on whether they (and RECs) deemed that the result or risk identified was medically actionable or individually meaningful, and whether they had obtained informed consent, so as to respect participants' right not to know (Downey et al. 2018). Other issues include whether there are resources to return individual results in an ethically responsible and feasible manner (Wong et al. 2018), and how false positive and negative results should be dealt with to minimize harm.

The 'research–practice' distinction and how we identify a proposed systematic data collection activity as one or the other are therefore ethically important, as they imply a shift in ethical commitments, standards, and requirements. As Case 4.2 and Case 4.3 suggest, and as is the practice in many settings, the classification of a systematic data collection activity as public health surveillance would typically mean that its ethical conduct was not contingent on prior independent ethics review. Conversely, determining that such an activity was research would mean there was a need to submit a formal protocol for REC review.

The normative line between research and practice in terms of types of data collection activities that need to or need not adhere to ethical safeguards such as informed consent and independent ethics review does not apply in every instance. For example, monitored emergency use of unproven interventions outside of clinical trials using a MEURI protocol requires informed consent from the recipients of the interventions and review by an REC or some other qualified ethics committee (see Table 4.2). Some scholars have also argued for ethical oversight of public health surveillance activities (Fairchild and Bayer 2004). Consistent with such thinking, the WHO endorses the establishment of an independent and impartial ethical oversight mechanism for public health surveillance systems (WHO 2017), while the PAHO states that surveillance activities, especially in the context of a public health emergency, should undergo some form of ethical oversight. WHO and PAHO are in agreement that ethical oversight of public health surveillance should not mimic approaches to research ethics oversight. The issue of ethical oversight of public health surveillance for OEP is discussed further below.

Regardless of the argument that public health surveillance should undergo some form of ethical oversight (or at least receive ethical guidance to ensure that they are conducted ethically), some methods or criteria are necessary to distinguish data collection activities carried out as research from non-research public health activities. This is to ensure adherence to any existing ethical requirements or regulations specific to these different activities, and to prevent their conflation, which could undermine the overall public health response to an OEP.

4.3   Locating the Distinction in the Primary Aim

Case 4.3 presents the question of whether a public health survey – deemed "necessary to inform planning for pandemic response" – should be considered a surveillance or a research activity. The common way to consider whether a public health activity ought to be classified as research or non-research is with reference to its primary aim. As mentioned, whereas the primary aim of a non-research activity is to prevent/reduce disease or improve health, the primary aim of research is to produce generalizable knowledge. In the context of health research, this has been defined as "theories, principles or relationships, or the accumulation of information on which they are based related to health, which can be corroborated by accepted scientific methods of observation and inference" (CIOMS 2016: p. xii). Basing the boundary between research and public health practice on "primary aim" may not make it clear in some cases. It could be argued that the primary aim of research in response to an OEP emergency is to produce generalizable knowledge precisely to develop the means of preventing, controlling or treating the disease that triggers the emergency (WHO 2015, see p. 23).

One way to make the demarcation clearer, and to better establish the primary aim of a proposed systematic data collection activity, is to consider the actors involved and their ethical duties qua the role they occupy, as well as the target beneficiaries of the activity (Taylor 2019). In brief, public health authorities and practitioners should design and conduct activities that promote and protect the collective health and safety of the community *within their jurisdiction* as their primary ethical duty. The potential use of the data to benefit communities outside of their jurisdiction is incidental or at most a secondary concern. In contrast, although researchers have an ethical duty to protect research participants, they also have a duty to ensure that their research holds the prospect of scientific and social value that will outweigh the risks and burdens participants undergo and the resources used. Scientific value and social value can be anticipated if the study is designed to produce knowledge that could benefit others in the future. Although the research process could benefit the participants and the knowledge produced could benefit the community from which they are drawn, the research should be designed to yield data and conclusions that could be generalized for use by those beyond the participating community to maximize scientific and social value.

Thus, it has been contended that "when the intent of the systematic public health data collection is to benefit those beyond the borders of the local jurisdiction" (Taylor 2019), it should be classified as research, even if the activity is conducted by a public health authority. Using "intent" or primary aim to distinguish research activities from non-research public health activities underscores the importance of research as a component of OEP response, as the scientific and social value of the knowledge it produces could apply to communities in other settings or to the global community.

4.3.1 Experimentation

For health interventions, an important consideration for what their primary aim ought to be is whether the intervention is experimental. Experimentation may be defined as "exposure of an individual or community to an activity not yet proven effective (i.e., not yet standard practice)" (Taylor 2019). Typically, non-research health interventions are standard measures with a proven history, or they are backed by scientific evidence of effectiveness in preventing disease and promoting individual health or the health of a community, and the intended known benefits outweigh the potential risks. Lack of or insufficient knowledge on the safety, effectiveness, and risk-benefit balance of a health intervention, and the potential of increasing certainty and understanding of these elements through research methods, are reasons for providing that intervention through research and its ethical safeguards. There are exceptional situations, however, where it is justified to provide experimental or unproven interventions outside a research context.

For instance, in the context of a public health emergency involving a novel pathogen where mortality is high and no proven treatments or prophylaxes exist, unproven interventions may be ethically provided on a case-by-case basis with the intent to save their lives or reduce their suffering through different non-research routes, such as off-label use, expanded access, and MEURI (Lysaght et al. 2022). Unproven interventions may also be made available to a given population to realize some public health goal through MEURI, or country-specific mechanisms for emergency use authorization to facilitate the availability and use of medical countermeasures ("unapproved medical products, or unapproved uses of approved medical products" – Krause and Gruber 2020). In general, the risk-benefit ratio should be favourable, and the conditions specific to each pathway that permit the ethical emergency use of an unproven intervention outside of clinical trials to benefit individuals or groups must be met.

Despite the availability of criteria to determine when its implementation is ethically justified, the MEURI framework can be seen as intrinsically complex. Although MEURI is defined as a non-research activity aimed at offering individuals or groups access to an unproven intervention that might benefit them, it calls for a contribution to the production of evidence through systematic collection, monitoring and dissemination of data (PAHO 2020:5). It is unclear what "production of evidence" means and what the evidence should be used for. Statements by WHO, such as "physicians overseeing MEURI have the same moral obligation to collect all scientifically relevant data on the safety and efficacy of the intervention as researchers overseeing a clinical trial" (WHO 2016, p. 36) may invite the view that "production of evidence" just is production of generalizable knowledge. If so, it suggests that MEURI is not that different from research, or is a form of observational research even though it is not a controlled clinical trial.

To set a clear boundary between monitored emergency use and research, one of the authors of this chapter (Voo) and his colleagues have argued that the aim of *monitored* emergency use is to protect the "safety of the patient(s) receiving the [unproven] intervention, with the *ancillary* benefit of collecting data on safety and effectiveness that could be used to inform clinical trial designs (e.g., dosage, patient population, outcome measures, etc)" (Lysaght et al. 2022, p. 336). The primary aim of data collection, monitoring and dissemination therefore is to directly benefit patients, even if it may (indirectly) contribute to the production of generalizable knowledge by providing data and evidence to inform the design of any subsequent clinical trial on the unproven intervention. In other words, "primary aim" provides a basis for distinguishing between research and monitored non-research emergency use of unproven interventions as an emergency care practice.

In sum, "primary aim" is, arguably, the central concept for classifying a systematic data collection activity as research or non-research, and this would be shaped by considerations such as the actors involved, the targeted beneficiaries, and whether the intervention is sufficiently proven. As "[e]mergency circumstances can lead to a blurring of limits between public health practice and research, both because of time constraints and because this limit is sometimes genuinely difficult to define" (Calain et al. 2009), it would be good practice to implement a third-party mechanism such as a national ethics committee – which appears to be used in Case 4.3 – to adjudicate on cases when there is confusion between the boundaries of research and practice (PAHO 2016).

4.4 Should Non-research Activities Be Conducted as Research Instead?

Scarce resources should be used to shore up the health system and non-research public health response during an emergency like COVID-19 to prevent disease, and loss of life and suffering. It is important to recognize, however, that research is also a key aspect of OEP response because certain critical questions can only be adequately answered by research methods (London and Kimmelman 2020). For example, randomized controlled trials to establish causal relationships between interventions and effects, in conjunction with other available knowledge, remain the primary way to prove or disprove the quality, safety and efficacy of medical products accepted by national regulatory agencies (Khadem Broojerdi et al. 2020). Thus, although many activities are legitimately conducted as non-research public health response during an emergency, it might be more justifiable to conduct certain activities as research to generate the evidence for proving or disproving hypotheses and propositions related to the infectious pathogen and prevention, control and treatment measures, which would also assist with preparedness for future similar emergency situations.

Public health and medical practitioners may however not initiate their activities as non-research activities despite good scientific and ethical reasons to conduct them as research instead. As suggested by Case 4.3, one reason why those involved in public health response may prefer to classify a data collection activity as surveillance and not research is the concern that research ethics oversight would impede the activity. Just as research should not unnecessarily impede emergency response, research ethics review should not unnecessarily impede research from being carried out, especially when it is a key component of emergency response.

Since 2008, the WHO has recommended various mechanisms to facilitate rapid and robust REC review for research during OEP emergencies (WHO 2010, 2020b) (see Chap. 6). Despite the implementation of rapid ethics review, public health practitioners may prefer to classify certain public health response activities as practice even though they could potentially be classified as research, because approval may still be stalled as a result of "substantive ethical concerns" by the REC (see Case 4.3). Or, because certain requirements, such as informed consent, would not be waived if the activity was classified as research, which could reduce their efficiency and effectiveness in achieving the public health goals. Another reason could be the belief that professional ethical expectations and best practices in public health are adequate for the protection of individual rights and interests and for the implementation of ethically sound practices (Lee 2019). Whether public health surveillance should undergo ethical oversight or would benefit from it, especially in an OEP emergency, may depend on the socio-cultural and political context in which the surveillance was conducted, and the agility and responsiveness of the oversight mechanism (Lee 2019).

In the bid to save individual patients or achieve some public health goal, such as reduction of infection incidence and disease burden, unproven interventions may be provided to specific individuals or populations through non-research pathways *at the same time* as they are being investigated in controlled clinical trials. As described in Case 4.4, this may cause confusion among non-research recipients, as well as research recipients and other stakeholders, about whether the intervention is proven or unproven. Also, it may pose complex questions; for example, what comprises sufficient evidence to justify monitored emergency use of unproven (medical) interventions outside clinical trials and who is responsible for establishing this? (PAHO 2020). (MEURI access should be provided on an exceptional basis during public health emergencies when proven interventions are absent or unsatisfactory; reasons should be given as to why a clinical trial cannot be initiated immediately instead; a favorable risk-benefit balance should be established by a qualified scientific advisory committee based on sufficient preliminary evidence of safety and efficacy; and that such use of unproven interventions do not unduly threaten other essential activities of prevention and management of a public health emergency, including research (WHO 2020a, b, Section 2.3)). In an age where the spread of information is amplified by social media and other digital platforms, permitting an experimental or investigational product to be used as a clinical or public health intervention may create a widespread perception that it is already a product with a proven safety and efficacy profile, which may inhibit the development of proven interventions through research (e.g. by increasing pressure for non-trial access and impeding trial recruitment). Hence, the risk of any type of unmonitored access (e.g. unmonitored "off-label" use or "compassionate use") to unproven interventions is that it may undermine the public health response to an OEP by contributing to the widespread uncontrolled use of unsafe or inefficacious unproven interventions (CIOMS 2016, Guideline 20), and result in more harm than good.

It is thus important to coordinate research and non-research public health activities so that the former is not undermined by the latter and can be effectively carried out to generate robust scientific evidence to inform and formulate responses to a public health threat. For example, regarding Case 4.3, one could argue that it is more effective and ethically appropriate for the public health survey to be conducted as research so that data collection is separated from the goal of active case detection.

Case 4.1 prompts the question of whether a prospective study on clinical outcomes of convalescent plasma treatment of COVID-19 patients should have been conducted instead of a retrospective study, or whether the convalescent plasmas should have been provided through a MEURI protocol (if requirements for MEURI had been met). The data collected through MEURI could have been used to support (or decide against) clinical trial initiation or to inform the design of such a trial. Providing the treatment through either research or MEURI would have required informed consent (unless there were good reasons against this, for example, patients did not have the capacity to give consent, in which case proxy consent could have been obtained). The extent to which patients should be given different information depending on whether they had the treatment through research or through MEURI is an interesting question.

For Case 4.4, the activity of providing the COVID-19 investigational vaccine to health-care workers (HCWs) could become part of the ongoing Phase II trial but its exclusion of those with current or previous SARS-CoV-2 infection would likely result in the non-participation of many HCWs, given their high risk of exposure to the virus and lack of vaccination. This raises the question of whether the trial's inclusion/exclusion criteria are justified, which is a matter of fair subject selection and depends primarily on the scientific aims of the trial (Emanuel et al. 2008). Given that the primary public health aim is to protect HCWs, however, it is important to consider whether it is ethically justifiable to provide the investigational vaccine to these workers through emergency use authorization (EUA). Whether it is justifiable to do so will depend on whether there is adequate interim trial data on its safety and efficacy (as determined by the relevant regulator) to support a favourable risk-benefit assessment. This is a key consideration for emergency use authorization, as is the potential of the medical product to prevent, diagnose or treat serious or life-threatening diseases or conditions (Singh and Upshur 2021). Again, an independent scientific and ethical oversight system could be involved in the assessment of the preliminary evidence as well as the risks and benefits of this non-research activity, so as to increase confidence that the vaccine is unlikely to cause net harm if offered under an emergency use authorization (in the US) or other form of monitored emergency use. In any case, "to minimize the risk that use of a vaccine under an EUA will interfere with long-term assessment of safety and efficacy in ongoing trials, it will be essential to continue to gather data about the vaccine even after it is made available under the EUA" (Krause and Gruber 2020).

4.5 Conclusion

Non-research activities, such as surveillance and emergency non-research use of unproven interventions, and research activities must be undertaken in an ethical manner as components of an OEP emergency response. To ensure this, it is important to identify these different activities on the basis of their primary aim, and to consider whether their implementation is in itself justifiable, based on their aims and the relevant ethical framework for each type of activity, and how they are coordinated as part of the larger collective activity of the OEP emergency response and management. How to make valid and defensible decisions on the type of response activity – whether research or non-research – is a complex question with different stakeholders involved in decision making. The aim is to ensure that research and non-research activities are appropriately distinguished, to ensure the proper coordination of such activities, and to increase trust and social accountability in OEP response.

Case 4.1: Use of Convalescent Plasma in Severely Ill COVID-19 Patients

This case study was written by members of the case study author group.

Keywords Boundaries between research, surveillance and clinical care; Data protection, access and sharing; Consent; Treatment repurposing; Emergency Use Authorisation

SARS-CoV-2 is responsible for a severe acute respiratory syndrome, which can cause death, particularly in more vulnerable people. The use of human convalescent plasma was considered as a potential treatment for COVID-19 (Casadevall and Pirofski 2020). While the United States Food and Drug Administration (FDA) had not approved its use for this purpose, it could be provided initially under the expanded use programme and subsequently through emergency use authorization as an investigational product (US Food and Drug Administration n.d.).

Human blood plasma has been used in restricted contexts to treat other viral diseases for which there are no established treatments. Previous studies with human plasma for treatment of Hantavirus cardiopulmonary syndrome showed that it appeared to be safe and able to reduce the fatality rate (Vial et al. 2015). Likewise, during the Ebola epidemic, the World Health Organization determined that, in the absence of other proven treatments, convalescent plasma could be authorized for "monitored emergency use" as an "unregistered and experimental intervention" when treating people with Ebola virus disease (WHO 2015).

In June 2020, a research ethics committee (REC) in South America, received a request to review a retrospective research protocol aimed at studying the use of convalescent plasma as a potential treatment for severe COVID-19 pneumonia. The researchers requested a waiver of informed consent for obtaining clinical data from the patients´ health records, because they considered it to be a protocol which aimed to evaluate a clinical practice, not a new therapeutic approach. At the time of the request, at a country level, there were a small number of ongoing clinical trials investigating the efficacy and safety of the use of convalescent plasma. In these cases, REC approval had been obtained prior to the administration of plasma, and both donors and recipients (or their representatives) of plasma had consented to participate in the trial. However, other institutions, such as the clinical centre requesting ethical approval for this retrospective trial, had adopted the administration of plasma as a clinical practice rather than an experimental intervention. Consequently they had not complied with guidance for the emergency use of unproven interventions outside of research, including the requirements for prior ethical review and informed consent (PAHO 2020). While plasma donors had given informed consent to donate their plasma for treatment, the patients who received convalescent plasma did not give informed consent to receive an unproven intervention, or consent for their clinical data to be used in research.The REC discussed the scope of its review: did the research encompass the administration of human plasma as an investigational product, or was it limited to the retrospective analysis of a clinical practice study?

Questions

1. In a pandemic, when emergency use authorizations are in place, are there morally significant differences between the prospective use of a medication under emergency use authorization, and a retrospective analysis of a clinical practice? What are the reasons for your views?
2. Given guidance for emergency use of unproven clinical interventions outside of trials, was it ethically acceptable for researchers to characterise their use of convalescent plasma as a clinical practice to be retrospectively evaluated? Why?
3. In this case should the REC focus on the ethical issues related to using an unproven intervention in a pandemic or on the request to access and analyse retrospective clinical data? Why?
4. Should a request for a waiver of informed consent to access the clinical data be granted in this situation? Why?

References

Casadevall A, and L. Pirofski. 2020. The convalescent sera option for containing COVID-19. *Journal of Clinical Investigation* 130(4): 1545–1548.

PAHO. 2020. Emergency use of unproven interventions outside of research ethics guidance for the COVID-19 pandemic. Pan American Health Organization. https://iris.paho.org/handle/10665.2/52429.

US Food and Drug Administration. n.d. Recommendations for investigational COVID-19 convalescent plasma. https://www.fda.gov/vaccines-blood-biologics/investigational-new-drug-ind-or-device-exemption-ide-process-cber/recommendations-investigational-covid-19-convalescent-plasma.

Vial, P.A., F. Valdivieso, M. Calvo, M.L. Rioseco, R. Riquelme, A. Araneda, V. Tomicic, J. Graf, L. Paredes, M. Florenzano, T. Bidart, A. Cuiza, C. Marco, B. Hjelle, C. Ye, D. Hanfelt-Goade, C. Vial, J.C. Rivera, I. Delgado, and G.J. Mertz (Hantavirus Study Group in Chile). 2015. A non-randomized multicentre trial of human immune plasma for treatment of hantavirus cardiopulmonary syndrome caused by Andes virus. *Antiviral Therapy* 20(4): 377–86.

WHO. 2015. Ethics of using convalescent whole blood and convalescent plasma during the Ebola epidemic: Interim guidance for ethics review committees, researchers, national health authorities and blood transfusion services. Geneva: World Health Organization. https://apps.who.int/iris/handle/10665/161912.

Case 4.2: COVID-19 Antibody-Testing Initiatives in a European Country

This case study was written by members of the case study author group.

Keywords Boundaries between research, surveillance and clinical care; Researcher roles and responsibilities; Researcher roles and responsibilities; Return of results

In 2020, a number of different programmes involving testing for SARS-CoV-2 antibodies were established in a European country.

A major biobank invited existing participants to provide blood samples, which would be tested for antibodies to provide data about the extent of previous infection in different parts of the country. This study was subject to standard ethical review processes. Results were not available to participants, on the grounds that this was a research programme and was established to study results at population and not individual level. As such it was not intending to offer a clinical or public health service, and at the time, the biobank took the view that feedback outside of the normal clinical setting would be of questionable value, and might even be harmful.

At the same time, health authorities in the country established an antibody-testing programme. This programme had a similar aim: to provide information on the prevalence of COVID-19 in different parts of the country and improve understanding of how the disease was spreading. This was not badged as research and did not undergo ethical review. Antibody testing was offered initially to health-care workers, and then to patients who were having a blood test for other purposes. Results regarding the presence or absence of COVID-19 antibodies were shared with participants.

Running alongside these two initiatives, a nationwide study of a sample of private households was run by the national statistical authority, in order to track levels of both current and past COVID-19 infection (using nose/throat swabs and blood samples respectively). This was considered to be research and was subject to ethical review. Participants were sent their results around a week after testing.

Questions
1. What implications does the pandemic context have for arguments for and against rapidly sharing antibody test results with participants in both research and public health surveillance contexts?
2. What are the ethically significant differences between the initiatives identified as research, and the initiative identified as public health surveillance?
3. Are these differences likely to be seen as relevant from the perspective of the participants taking part? Why?
4. Could these differences in approach be justified by technical reasons, for example relating to the likely accuracy of individual results? If so, how could this be handled so that participants feel informed and respected?

Case 4.3: Competing Priorities Under Pressure: Government Collaboration with Academic Institutions

This case study was written by members of the case study author group.

Keywords Boundaries between research, surveillance and clinical care; Researcher roles and responsibilities; Ethical review; Safety and participant protection; Consent; Privacy and confidentiality

Political leaders and the public health agency in a small lower-middle-income country implemented urgent measures to contain the acute spread of COVID-19, including a number of social and economic restrictions. These measures were developed in consultation with a team comprising academic, political and public health leaders. In response to an increase in COVID-19 cases, the government decided that a door-to-door household survey of COVID-19-related symptoms and behaviours (e.g. adherence to non-pharmaceutical interventions) was necessary to inform planning for the pandemic response. The president of the country announced the initiative on television.

A member of the survey team contacted the national research ethics committee to discuss whether research ethics review should be required. The survey team had expressed differing views in internal discussions, with the majority holding that the survey did not require review because they conceptualized the project as a government-endorsed public health surveillance activity.

Moreover, the survey team viewed the research ethics review process as having the potential to unnecessarily delay the collection of information urgently needed to contain the pandemic. The team held these views even though the ethics committee had developed and publicized a rapid review process 10 months previously, at the start of the COVID-19 outbreak. The process had facilitated a number of reviews on an accelerated schedule for COVID-19-specific research, as well as on-going studies requiring amendments to accommodate increased safety measures to minimize risks of SARS-CoV-2 transmission. During the course of the rapid review programme, some proposals had been quickly actioned; others had taken longer, owing to both substantive ethical concerns and operational challenges for the ethics committee.

As the survey team had not submitted a protocol to the ethics committee, the proposed procedures for informed consent and other safeguards to protect the rights and well-being of participants were not clear. The conversation between the survey team and ethics committee revealed a number of potential ethical concerns that merited consideration. There was a possibility that the survey data collectors would be asked to inform public health authorities if they observed survey participants showing signs of COVID-19. However, the data collection protocol for the survey did not include prior informed consent, raising the possibility that persons answering a knock at the door might be reported to government health authorities, an action that could result in involuntary isolation at a quarantine facility. Political leaders emphasized publicly that the door-to-door contact would provide the opportunity to identify people with suspected COVID-19 who potentially posed risks to others or were at risk themselves. In the context of the pandemic the survey would therefore not just collect data, but also potentially act as a mechanism to limit spread of infection, with a consequent impact on individual liberties.

Further, law enforcement and military personnel were assisting in both the humanitarian and enforcement components of the public health response to the pandemic. Although soldiers and police would not be conducting the survey directly, they would provide security for the survey team and might be visible while survey data were being collected.

Questions
1. Should a survey of this type be considered a public health surveillance measure or a research study? Why?
2. How should research ethics committees respond when they have been informally notified of a survey that raises concerns but are advised that it will not be submitted for review due to the urgency of conducting it?
3. How should public health surveillance activities be better coordinated with (or differentiated from) research efforts in the context of a pandemic, especially when the priorities of research ethics committees (e.g. ensuring the ethical conduct of research, including protecting the interests of participants) and public health authorities (e.g. containing spread of infection) might conflict?
4. What concerns might you have regarding protections for survey respondents in this scenario? Would the concerns change if the survey was seen as purely a public health initiative to inform pandemic responses without a research component?

Case 4.4: Vaccine Research or Rollout?

This case study was written by members of the case study author group.

Keywords Boundaries between research, surveillance and clinical care; Research design and adaption; Risk/benefit analysis; Vulnerability and inclusion; Vaccines

In a South American country, the government was making an effort to combat the advance of COVID-19 in its territory. It was the first country to carry out a clinical investigation of a new recombinant vaccine for the prevention of COVID-19.

A clinical trial aimed to assess the vaccine's efficacy against the numerous variants of the SARS-CoV-2 virus circulating in the country. The vaccine had received emergency use authorization and the Phase II study was conducted in the main cities of the country. To reach recruitment objectives, the population was called on to volunteer for the study. All participants would receive free medical care, as well as private medical insurance, and reimbursement for food, transportation and medical consultations. In light of these benefits, and especially the potential of gaining immunity to COVID-19, many people volunteered to join the trial. An exclusion criterion for the trial was a current or previous infection with SARS-CoV-2.

While the trial was being conducted, health authorities also authorized the rollout of the new recombinant vaccine to health-care workers. The aim was to provide protection against infection to health-care staff who were frequently in contact with COVID-19 patients and therefore playing a key role in the fight against the pandemic. In contrast to the trial participants, health-care workers were offered the vaccination even if they had previously been infected with SARS-CoV-2.

This led to questions among health-care workers about whether they were participating in a trial or being offered the vaccination as a preventative intervention. Some health-care workers were grateful to be the first to receive the vaccine, since they were most likely to be exposed to infection. However, others thought the opposite, noting that as the Phase II trial was still being conducted, the vaccine's safety and efficacy were not proven. For this reason, they opposed receiving the experimental vaccine.

Questions
1. During the Phase II trial is it acceptable to present this COVID-19 vaccine as an intervention to protect health-care workers? Why?
2. If exposure to SARS-CoV-2 infection is an exclusion criterion in the Phase II trial, is it ethical to offer the vaccine to health-care workers (who are likely to have been exposed to infection)? Why?
3. Are there conditions under which it would be ethically acceptable for a COVID-19 vaccine trial to have inclusion criteria restricting recruitment to health-care workers? Why?
4. Is it acceptable for health professionals to refuse the experimental vaccine, given the contact they have with vulnerable patients? Why?

References

Barrett, D.H., L.W. Ortmann, N. Brown, B.R. DeCausey, C. Saenz, and A. Dawson. 2016. Public health research. In *Public health ethics: Cases spanning the globe*, ed. D.H. Barrett, L. Ortmann, A. Dawson, C. Saenz, A. Reis, and G. Bolan. Cham: Springer.

Beauchamp, T.L., and Y. Saghai. 2012. The historical foundations of the research-practice distinction in bioethics. *Theoretical Medicine and Bioethics* 33(1): 45–56.

Calain, P., N. Fiore, M. Poncin, and S.A. Hurst. 2009. Research ethics and international epidemic response: The case of Ebola and Marburg Hemorrhagic fevers. *Public Health Ethics* 2(1): 7–29.

CIOMS. 2016. *International ethical guidelines for health-related research involving humans*. Geneva: Council of International Organizations of Medical Sciences. https://cioms.ch/wp-content/uploads/2017/01/WEB-CIOMS-EthicalGuidelines.pdf.

Doerr, M., and J.K. Wagner. 2020. Research ethics in a pandemic: Considerations for the use of research infrastructure and resources for public health activities. *Journal of Law and the Biosciences* 7(1): lsaa028.

Downey, A.S. E.R., Busta, M. Mancher, and J.R. Botkin. 2018. *Principles for the return of individual research results: Ethical and societal considerations*. Washington: National Academies Press. https://www.ncbi.nlm.nih.gov/books/NBK525079/.

Emanuel, E.J., D. Wendler, and C. Grady. 2008. What makes clinical research ethical? In *The Oxford textbook of clinical research ethics*, ed. E. Emanuel, C. Grady, R. Crouch, R. Lie, F. Miller, and D. Wendler. New York: Oxford University Press.

Fairchild, A.L., and R. Bayer. 2004. Ethics and the conduct of public health surveillance. *Science* 303(5658): 631–632.

Khadem Broojerdi, A., H. Baran Sillo, R. Ostad Ali Dehaghi, M. Ward, M. Refaat, and J. Parry. 2020. The World Health Organization global benchmarking tool an instrument to strengthen medical products regulation and promote universal health coverage. *Frontiers in Medicine* 7: 457. https://doi.org/10.3389/fmed.2020.00457.

Krause, P.R., and M.F. Gruber. 2020. Emergency use authorization of Covid vaccines – Safety and efficacy follow-up considerations. *New England Journal of Medicine* 383: e107. https://doi.org/10.1056/NEJMp2031373.

Lee, L.M. 2019. Public health surveillance: Ethical considerations. In *The Oxford handbook of public health ethics*, ed. A.C. Mastroianni, J.P. Kahn, and N.E. Kass. New York: Oxford University Press.

Lockwood, A., and H. Walters. 2018. Making the most of public health research. *Journal of Public Health* 40(4): 673–674.

London, A.J., and J. Kimmelman. 2020. Against pandemic research exceptionalism. *Science* 368 (6490): 476–477.

Lysaght, T., G.O. Schaefer, T.C. Voo, et al. 2022. Professional oversight of emergency-use interventions and monitoring systems: Ethical guidance from the Singapore experience of COVID-19. *Journal of Bioethical Inquiry* 19: 327–339. https://doi.org/10.1007/s11673-022-10171-1.

Mastroleo, I., and F. Holzer. 2020. New non-validated practice: An enhanced definition of innovative practice for medicine. *Law, Innovation and Technology* 12(2): 318–346.

National Commission for the Protection of Human Subjects of Biomedical and Behavioral Research (National Commission). 1978. *The Belmont report: Ethical principles and guidelines for the protection of human subjects of research*. Washington: US Department of Health, Education and Welfare.

Otto, J.L., M. Holodniy, and R.F. DeFraites. 2014. Public health practice is not research. *American Journal of Public Health* 104(4): 596–602.

PAHO. 2016. *Zika Ethics Consultation: Ethics guidance on key issues raised by the outbreak*. Pan American Health Organization. https://iris.paho.org/handle/10665.2/28425.

————. 2020. *Emergency use of unproven interventions outside of research ethics guidance for the COVID-19 pandemic*. Pan American Health Organization. https://iris.paho.org/handle/10665.2/52429.

Schaefer, G.O., G. Laurie, S. Menon, A.V. Campbell, and T.C. Voo. 2020. Clarifying how to deploy the public interest criterion in consent waivers for health data and tissue research. *BMC Medical Ethics* 21(1): 23.

Singh, J.A., and R.E.G. Upshur. 2021. The granting of emergency use designation to COVID-19 candidate vaccines: Implications for COVID-19 vaccine trials. *The Lancet Infectious Diseases* 21(4): e103–e109.

Taylor, H.A. 2019. Framing public health research ethics. In *The Oxford handbook of public health ethics*, ed. A.C. Mastroianni, J.P. Kahn, and N.E. Kass, 331–341. New York: Oxford University Press.

Verweij, M., and A. Dawson. 2007. The meaning of 'public' in 'public health'. In *Ethics, prevention, and public health*, ed. A. Dawson and M. Verweij, 13–29. Oxford: Clarendon Press.

WHO. 2010. *Research ethics in international epidemic response: WHO technical consultation*. Meeting report. Geneva, 10–11 June 2009. World Health Organization. https://apps.who.int/iris/handle/10665/70739.

————. 2015. *Ethics in epidemics, emergencies and disasters: Research, surveillance and patient care: Training manual*. Geneva: World Health Organization. https://apps.who.int/iris/handle/10665/196326.

————. 2016. *Guidance for managing ethical issues in infectious disease outbreaks*. Geneva: World Health Organization. https://apps.who.int/iris/handle/10665/250580.

————. 2017. *WHO guidelines on ethical issues in public health surveillance*. Geneva: World Health Organization. https://apps.who.int/iris/bitstream/handle/10665/255721/9789241512657-eng.pdf.

————. 2018. *Consultation on monitored emergency use of unregistered and investigational interventions for Ebola Virus Disease (EVD)*. Meeting report. May 17. World Health Organization. https://www.who.int/publications/m/item/consultation-on-monitored-emergency-use-of-unregistered-and-investigational-interventions-for-ebola-virus-disease.

————. 2020a. *Ethical standards for research during public health emergencies: Distilling existing guidance to support COVID-19 RandD*. Policy Brief. Geneva: World Health Organization. https://apps.who.int/iris/handle/10665/331507

————. 2020b. *Guidance for research ethics committees for rapid review of research during public health emergencies*. May 28. Geneva: World Health Organization. https://www.who.int/publications/i/item/9789240006218.

Wong, C.A., A.F. Hernandez, and R.M. Califf. 2018. Return of research results to study participants: Uncharted and untested. *Journal of the American Medical Association* 320(5): 435–436.

Chapter 5
Adapting and Adaptive Research

Maxwell J. Smith

Abstract Research conducted during epidemics may warrant adaptations or adaptive designs owing to practical constraints, time pressures, uncertainty, the importance of flexibility, and the potential for research to detract from epidemic response. Adapting research entails choosing different research designs or methods if research goals, contexts or constraints justify or require a different approach. Adaptive research, by contrast, is a type of research that prospectively plans for modifications after research has been initiated, while maintaining the validity and integrity of the research. While adaptation and adaptive designs introduce an important degree of flexibility to research conducted during epidemics and help to address research objectives and constraints, adaptation and adaptive designs require close ethical scrutiny and are no different from other research in that they are expected to align with universally accepted ethical standards. Important ethical questions exist regarding the conditions that justify adaptations to research, the kinds of adaptive research designs that can be ethically justified, and how ethics review bodies ought to evaluate such novel approaches to research in epidemic contexts. The five cases included in this chapter prompt reflection on the ethical considerations and implications of adapting research in response to epidemic-related risks and the public health measures deployed in response to those risks, as well as the ethical implications of *not* adapting research in such contexts. These cases also highlight ethical questions and issues arising during the conduct of adaptive trials, including when treatments under study, treatment doses, sample size, and other study features are reviewed in response to evolving evidence. This chapter invites reflection on these key ethical dimensions when considering adaptive designs and adaptations to standard research procedures during epidemics. What these cases make clear is that adaptive designs and adaptations to research do not reduce the need for rigorous scientific evaluation and adherence to universal ethical standards, and must be explicitly ethically justified and reviewed through transparent and inclusive processes.

M. J. Smith (✉)
Faculty of Health Sciences, Western University, London, ON, Canada
e-mail: maxwell.smith@uwo.ca

© PAHO and Editors 2024
S. Bull et al. (eds.), *Research Ethics in Epidemics and Pandemics: A Casebook*,
Public Health Ethics Analysis 8, https://doi.org/10.1007/978-3-031-41804-4_5

Keywords COVID-19 pandemic · Research ethics · Public health emergencies · Research design and adaptation · Privacy and confidentiality · Data protection · Access and sharing · Vulnerability and inclusion · Digital and remote healthcare and research · Risk/benefit analysis · Pausing and halting research · Community engagement and participatory processes · Resource allocation · Non COVID-19 research · Safety and participant protection · Access to experimental treatments · Consent · Ethical review · Researcher roles and responsibilities · Research priority setting · Research publication ethics · Social and scientific value

5.1   Introduction

Conducting research during epidemics is of critical importance. Yet, owing to practical constraints, time pressures, uncertainty, the importance of flexibility, and the potential for research to detract from epidemic response, the ways in which research is conducted in this context may warrant adaptations or adaptive designs.

Adapting research entails choosing different research designs or methods if research goals, contexts or constraints warrant or necessitate a different approach. For example, as Case 5.1 illustrates, researchers who initially planned to conduct research involving face-to-face interaction to generate data may have to instead adopt remote forms of data generation, owing to epidemic-related risks and measures that restrict mobility. And as Cases 5.2 and 5.3 highlight, *failing* to adapt and instead pausing research, given the challenges presented by an epidemic, may have negative impacts on research participants.

Adaptive research, by contrast, is a type of research that prospectively plans for modifications after research has been initiated, while maintaining the validity and integrity of the research (Mahajan and Gupta 2010). For instance, research may be designed with a plan to revisit the treatments under study, the treatment doses, the sample size, and so forth (Pallmann et al. 2018). Case 5.4 provides a nice example of adaptive research via the RECOVERY Trial, wherein study arms were added when there was reason to believe an intervention offered a benefit or removed when sufficient data had been collected to establish that an intervention was associated with a lack of benefit. And Case 5.5 highlights the possible implications of *not* using an adaptive research design in the context of evolving scientific evidence.

While adaptation and adaptive designs introduce an important degree of flexibility to research conducted during epidemics and help to address research objectives and constraints, this does not mean that "anything goes". Scientific rigour and validity, in addition to adherence to universally accepted ethical standards, remain essential. Consequently, important ethical questions exist regarding the conditions that justify adaptations to research, the kinds of adaptive research designs that can be ethically justified, and how ethics review bodies ought to evaluate such novel approaches to research in epidemic contexts.

5.2 Adapting – Not Deviating From – Scientific and Ethical Standards for Research

If epidemic contexts sometimes warrant or necessitate that research be adapted or adaptive, does this mean that exceptions should be made to the scientific and ethical standards that otherwise govern research? Such standards include both scientific standards, such as those commonly used for participant selection, sample size estimation and sample size allocation, as well as ethical standards, such as those governing the ethics review and informed consent processes (see Chap. 6).

London and Kimmelman argue that the challenges that rigorous scientific methods are designed to address do not disappear during public health emergencies like epidemics, nor do researchers' obligations to align the conduct of their research with the public interest or to protect the interests of research participants, both of which are advanced by research ethics standards and regulations (London and Kimmelman 2020). In other words, they argue that "the moral mission of research remains the same: to reduce uncertainty and enable caregivers, health systems, and policy-makers to better address individual and public health" (p. 476). Consequently, while accepted ethical and scientific standards should be interpreted in light of, and adapted in response to, particular circumstances and contexts in epidemics, the aim must still be to generate the best possible evidence about important questions. Adaptive research designs and adaptations to research therefore do not sidestep the need for rigorous scientific evaluation and adherence to universal ethical standards and must be explicitly ethically justified and reviewed through transparent and inclusive processes.

The evidence base can evolve rapidly during an epidemic. Researchers and those charged with reviewing ongoing studies (e.g. research ethics committees) therefore have a responsibility to monitor emerging evidence from other research initiatives, review the implications for the studies they are leading or overseeing and decide whether those studies should be continued, modified, suspended or cancelled, in order that they continue only if they have scientific and social value and so that people are not asked to participate in research that is no longer likely to produce meaningful results or which poses risks without the prospect of benefit (PAHO 2020). In other words, the justification and ethical acceptability of research can vary throughout its duration as a result of rapidly evolving evidence. Decision-making in this context can be particularly challenging as evidence may be uncertain or contested. Consequently, the intervals at which studies are reviewed and report to research oversight bodies ought to be shorter and more frequent during an epidemic. Researchers should also develop plans that account for how their study might be affected by new evidence or adapted in response to it.

The Nuffield Council on Bioethics (Nuffield Council on Bioethics 2020, pxxii) raises two key questions that they suggest ought to be asked when considering adaptations to standard research procedures during public health emergencies, like epidemics:

Is this the right study for this location and this population/subpopulation? Who has been involved in identifying and characterising the problem that the research seeks to answer? Will local populations benefit from any positive findings?

Is this the right design for this location and this population? How have local needs, concerns or preferences been taken into account?

Following on from these questions, the Council offers two recommendations:

> Study protocols should be developed with the input of local communities before being finalised, in order to ensure that proposed procedures are acceptable to communities, as well as meeting ethical requirements. Even in multi-site trials, there will be elements that can and should be operationalised differently in different sites in response to engagement and feedback.
>
> Any exclusion criteria from studies should be clearly justified with reference to the risks and benefits for the group in question, in this context, rather than an automatic exclusion of 'vulnerable groups'.

5.3   Adaptive Clinical Trials

Adaptive clinical trials (ACTs) are a particularly salient type of adaptive research that may be considered during epidemics. For instance, the magnitude and high case fatality rate of the 2014–2016 Ebola virus disease (EVD) epidemic in West Africa prompted calls for the accelerated evaluation and development of investigational therapeutic interventions that had shown promising results in the laboratory and in animal models. In response, a World Health Organization ethics advisory panel concluded that it was ethical to offer investigational agents with the intent to treat those suffering from EVD, and that a moral duty existed to evaluate these interventions in the best possible clinical studies (WHO 2014); however, it was unclear what the ethical requirements were for the appropriate design of such investigations. Proponents of placebo-controlled randomized controlled trials (RCTs), for instance, argued that these designs ought to be used as they were best able to generate robust, statistically valid evidence about safety and efficacy, which could be used to ensure all patients receiving treatment after the trial received the safest and most effective intervention (Joffe 2014). On the other hand, proponents of ACTs argued that ACTs would be preferable as they better allow for emerging, accumulated data to be used to rapidly identify and deploy beneficial new therapies to improve outcomes among trial participants (Adebamowo et al. 2014).

The principal argument favouring the conduct of placebo-controlled RCTs in the context of epidemics and other public health emergencies is that one ought to collect the best possible evidence in order to develop the safest and most effective intervention, and that a placebo-controlled RCT is the most appropriate, and perhaps morally obligatory, method of achieving this goal. The principal argument favouring the conduct of ACTs in the context of epidemics and other public health emergencies is that, owing to the severity and urgency present during epidemics, in addition to the higher fatality rates associated with conventional, supportive care in the absence of effective therapies, one should give greater weight to the well-being of the patients affected and therefore favour ACTs, given their ability to adapt to emerging evidence of treatment safety and efficacy (Singh 2023). Table 5.1 outlines the key relative merits of RCTs and ACTs, as well as the ethical considerations regarding each, in order to elucidate the potential value, as well as the potential pitfalls, of conducting ACTs during an epidemic.

Table 5.1 Relative merits of RCTs and ACTs and ethical considerations regarding each type of trial

	Placebo-controlled randomized controlled trials	Adaptive clinical trials
Principal design advantages	• Design with which regulators may be most familiar/comfortable • Considered by many the "gold standard" design for generating statistically valid evidence about safety and efficacy • Given background assumptions, hypotheses follow deductively from results, leading to high internal validity • Randomization aims to control for confounding factors and ensure that individuals receiving investigational agents do not systematically differ from individuals receiving only conventional therapy • Randomization effectively blinds investigators and controls for selection bias	• Aims to balance the production of scientifically valid knowledge with the intent to alleviate trial participants' suffering • Can limit participant exposure to unnecessary/ineffective interventions (e.g. by dropping an experimental arm during a trial) • Perhaps most appropriate in desperate, life-threatening situations, where the risk to the individual patient is greatest, as information generated earlier in the trial may inform the allocation of investigational therapies later in the trial • Flexibility in modifying study parameters during the study, given evolution of epidemic • If new interventions are available, ability to add promising interventions and drop ineffective interventions without restarting trial • External information can be incorporated into the study while in progress
Principal design disadvantages	• In order to make deductive inferences, the scope of the inference is limited, affecting external validity • Many confounding effects will invariably exist in the context of epidemics, which may challenge the ability to make valid inferences from trial populations to target populations • Randomization might not be feasible in the context of healthcare systems that are non-existent or under great strain • Randomization tends to aim to gather robust and well-controlled information but largely ignores immediate responses, which may undermine trust in the epidemic response	• Regulators not as familiar or comfortable with design • Design not as well understood or accepted in scientific community • Without doing an RCT, claims that any interventions are safe and efficacious may be seen to carry less weight • Potential for insufficient top-down financial and motivational support from R&D organizations • Potential requirement of additional time for planning • Lack of blinding (when not used) may increase response bias • Lack of a concurrent control group (when not used) may confound efforts to reach valid inferences about the investigational agent's safety and efficacy

(continued)

Table 5.1 (continued)

	Placebo-controlled randomized controlled trials	Adaptive clinical trials
		• If unblinded, information could be leaked, which could jeopardize trial recruitment or credibility in future trials • Owing to its flexible nature, design adaptations may be challenging for conventional statistical methodology for analysis
• Who may be most benefited by design?	• Individuals affected by an epidemic disease following the trial (including in future epidemics), if therapy is found to be efficacious • Manufacturers of the therapeutic agent, as RCTs provide perhaps the best pathway for drug development and licensure	• Trial participants, who are treated as effectively as possible given current and emergent evidence • Affected communities, as the rapid identification and deployment of beneficial therapeutic agents could in turn curb the spread and impact of disease
• Who may be most burdened by design?	• If the investigational therapeutic agent is efficacious, those randomized to control group • If the investigational therapeutic agent is harmful, those randomized to experimental group	• Trial participants enrolled earlier in the study, owing to adaptive nature • Manufacturers of therapeutic agent, insofar as additional trials may be required following ACTs in order to develop and license agent for broader use
• Additional ethical considerations in the context of epidemics	• RCTs are only ethical when equipoise exists, which could be undermined when conventional care offers little benefit for diseases with high rates of mortality (e.g., emerging evidence of efficacy for an investigational agent may more quickly appear 'better' than conventional care). • Scientific validity may be distorted if potential participants fabricate inclusion criteria because of desperation	• ACTs attempt to provide a compromise between data generation on safety and efficacy that is used to inform future decisions, and utilizing accumulated data to alleviate suffering for current patients • Criteria ought to be developed to guide the level and scope of design adaptation

5.4 Adapting Research to Epidemic Contexts

The ethical appropriateness of any research design should to some extent be informed by the context in which the research is to be conducted (Pullman and Wang 2001). That is, it has been argued that methodological orthodoxy ought to be eschewed in order to critically consider the research context, background information, risks of the research and the most appropriate means of answering specific research questions and achieving stated goals (Pullman and Wang 2001; Cartwright 2007; Ezeome and Simon 2010). Appreciating the motivations for and principal objectives of conducting clinical trials in the context of an epidemic may, at least in part, be instructive of which trial design ought to be favoured, and whether adaptations are ethically justifiable (if not ethically obligatory). For example, while not necessarily mutually exclusive, there were at least two central objectives that were advanced in relation to conducting trials in the midst of the EVD epidemic: (1) to aid the current humanitarian response and to make potential therapies rapidly available in order to save as many lives as soon as possible; and (2) to generate the most robust, scientifically valid data that would lead to the development of a licensed product that could, in turn, be used to ensure the safest and most efficacious intervention was available for patients receiving treatment following the conclusion of the trial. Preference for either trial design in the context of an epidemic may therefore be dependent, at least in part, on which objective is considered the priority.

A World Health Organization ethics advisory panel argued that investigational therapeutic options should not divert resources or attention from the public health measures, which they claim ought to remain the priority in an epidemic response (WHO 2014). In 2022, WHO argued that "emergency use of interventions for which there is insufficient evidence of safety or efficacy for regular use in health systems is ethically permissible outside clinical trials or other research contexts, if the primary aim is clinical benefit for individual people or groups or benefit for populations, and if such use during public health emergencies complies with a sound ethical framework that ensures adequate justification, ethical and regulatory oversight, consent process and contribution to evidence" (WHO 2022). Others have warned that research conducted during an epidemic or other public health emergency could have the effect of encouraging the modification of public health priorities, perhaps from providing a humanitarian response to the rigorous collection of data (Ezeome and Simon 2010). As such, if substantial resources are to be invested to conduct a trial during an epidemic, then there is a strong argument to be made that a moral responsibility exists to use those resources in such a way that they benefit those affected by the epidemic and curb the further spread of the epidemic. While any trial design has the potential to direct attention away from the immediate epidemic response, it appears that ACTs may be more congruent with the immediate epidemic response, although, placebo-controlled RCTs could be designed in a way that makes them align better with the advantages of ACTs. This could be accomplished, for example, by utilizing stepped-wedge RCTs, which involve random and sequential crossover of clusters of participants from a control arm (or arms) to the experimental arm (or arms) until all clusters have been exposed to the experimental intervention (Hemming et al. 2014). Or, placebo-controlled RCTs could utilize data safety and monitoring boards, who are charged with reviewing interim data and implementing early stopping rules based on safety and/or efficacy thresholds.

It is important to acknowledge that, for any research conducted in the context of an epidemic, the ability of participants to provide informed consent may become compromised or the consent process may become less feasible (see Chap. 9). This may be due to participants' lack of mental or physical capacity in such dire circumstances, a lack of local health-care workers available to recruit participants, and/or a strong therapeutic misconception undermining participants' abilities to appreciate the clinician's dual role as researcher and health-care worker (Pullman and Wang 2001; Ezeome and Simon 2010; Adebamowo et al. 2014; Kass 2014; Tangwa 2014). As such, some argue that every effort must be made to provide the most effective treatment to every trial participant, given current information, and that ACTs attempt to accomplish this very task while still ensuring that research objectives can be pursued (Pullman and Wang 2001). The dire circumstances and the prospect of inevitable therapeutic misconception during the EVD epidemic led some to argue that entering West Africa with the aim of doing anything other than saving the lives of those affected by EVD and curbing the spread of the epidemic would be morally irresponsible (Tangwa 2014). This sentiment, if it is to be balanced with the motivation and need to collect crucial evidence about the safety and efficacy of investigational therapies, may be supportive of adopting an adaptive design.

5.5   Conclusion

Epidemics should prompt researchers to evaluate whether their research ought to be adapted or whether adaptive designs might be appropriate and ethically justified. What is clear is that adaptive designs and adaptations to research do not obviate the need for rigorous scientific evaluation and adherence to universal ethical standards, and must be explicitly ethically justified and reviewed through transparent and inclusive processes. Involving the voices of local, affected communities in research planning, design and oversight remains crucial. Engaging local communities in such aspects of the research may foster trust in the research and epidemic response and better ensure local values and customs are both respected and represented (Modlin et al. 2023). Consequently, the input of those affected by an epidemic and who may be impacted by any research conducted ought to be considered of the utmost importance in responding to the question of whether and how research might be adapted.

Case 5.1: Adapting Face-to-Face Interviews to Respect Infection Control Measures

This case study was written by members of the case study author group.

Keywords Research design and adaptation; Privacy and confidentiality; Data protection, access and sharing; Vulnerability and inclusion; Qualitative research; Digital and remote healthcare and research; Researcher safety

In April 2020, a non-governmental research group decided to conduct a prospective study in a Caribbean country to inform and orient governmental strategies for controlling the spread of COVID-19. The initial study design incorporated a population-based survey that would seek to recruit participants with a variety of demographic characteristics. The research team would evaluate their knowledge of COVID-19 and try to understand the reasons behind their non-compliance with the main infection-control measures proposed by the health authorities, including wearing masks and practising hand hygiene and physical distancing.

The researchers planned to undertake face-to-face interviews with study participants after obtaining informed consent. This strategy involved visiting crowded places like markets and talking to several people for about 30–45 min each. To meet the study objectives, it would be necessary to interview people who did not follow the infection-control measures. However, this mode of data collection might also prevent researchers from complying with physical distancing requirements themselves, raising concerns about the safety of both the researchers and the participants. The study schedule might also be affected by lockdown periods, and if it was, it would be unclear when research could restart.

To address the research team's concerns about exposing researchers and participants to infection, the principal investigator proposed to revise the protocol and conduct an online survey instead, using a structured four-page questionnaire. Before going through the questionnaire, participants would be shown a page presenting the research team, the study objectives, and the main ethical obligations, including data confidentiality. At the end of this page, the respondent would be invited to complete the questionnaire. Only two team members, a data officer responsible for the quality control and data analysis and the principal investigator, would access the data collected, which would be anonymized. The research team planned to advertise the questionnaire to the study population via organized groups: church groups, scientific communities and neighbourhood committees.

However, the choice of an online survey, while reducing risks to participants and researchers, has some limitations. For example, the study sample would not be representative of the population, because some people did not have access to the internet or did not have enough knowledge of online platforms to take part in an internet-based study. Translating the questionnaire into a local language at the request of the research ethics committee would make it accessible to members of the study population who did not read an official European language. However, the exclusion of the part of the population who could not read at all would still stand.

The face-to-face survey would have given the research team the opportunity to include this population by reading the questionnaire to them. Additionally, an online survey would provide less opportunity to investigate the causes of non-compliance in depth, with implications for the interpretation of results. During the administration of the pilot questionnaire, concerns also arose about the reluctance of the respondents to reveal their economic status in an online survey.

Questions

1. Should national research ethics committees provide guidance about ethical issues and considerations arising during research in pandemics in order to help researchers develop appropriate study designs? Why? If such guidance is produced, what issues should it address?
2. In a pandemic, how should researchers and public health officials address tensions between the need to conduct relevant research rapidly and the need to respect participants' safety and integrity?
3. The COVID-19 pandemic has exacerbated existing social disparities and created new vulnerabilities. How should these be addressed during the design and implementation of a research project during the pandemic?

Case 5.2: A Community-Based Intervention for Indigenous Older Persons with Mild to Moderate Dementia

This case study was written by members of the case study author group.

Keywords Research design and adaptation; Risk/benefit analysis; Pausing and halting research; Vulnerability and inclusion; Community engagement and participatory processes; Resource allocation; Non COVID-19 research

Cognitive stimulation therapy (CST) has been shown to improve cognition, mood and quality of life in adults with mild to moderate dementia. Delivered in twice-weekly group sessions over a 7-week period, it is considered safe and is used internationally, including in some regions of a country in the Western Pacific.

Within this country, gerontology researchers partnered with an indigenous community in a relatively remote area, where CST was not available. Working together over time to develop trust and mutual understanding, the researchers and the community adapted the programme for culturally appropriate delivery, by using the indigenous language. The researchers and community commenced a trial to determine its effectiveness in this population, with several rounds of participant recruitment planned. The indigenous community was instrumental in recruiting participants and facilitating the trial.

The first round of the intervention study, involving ten participants, was completed before COVID-19 emerged. It demonstrated significant improvements over baseline measurements for both cognitive function and mood and was well received by the participants and their wider family networks. The second group of participants had been recruited, and baseline measurements completed, when the country was placed under long lockdown restrictions. In line with public health directives, the trial was put on hold.

After several months, lockdown restrictions were eased, and some normal activities were able to resume. The dementia of some recruited participants had progressed from moderate to severe in the intervening period, so they were no longer eligible to participate in the study. Baseline measurements of dementia, cognition, mood and quality of life would have to be completed again for the remaining participants, and new potential participants identified. Researchers and their indigenous partners considered whether to proceed with the recruited participants and decided against doing so. Considerations included the burden of running the programme upon local community workers, who were stretched by COVID-19-response work, and concern that the pandemic increased the risks of participating and that this would cause anxiety among participants and their families. A decision was made to re-attempt recruitment at a later date when the outlook regarding COVID-19 rates in the community and the responses of public health were more certain.

During the months following that decision, localized lockdowns occurred outside the study area, highlighting the on-going risk of transmission of the SARS-CoV-2 virus. Community workers who would be involved in recruitment and in delivering the intervention were managing increased demands for their services. The researchers had time-limited funding to develop relationships with indigenous health providers in five more sites, with a view to rolling out the adapted programme more widely. However, those efforts depended upon the trial being completed.

Once COVID-19 was well contained nationally, there were no government restrictions preventing the trial, and there was no evidence of the virus circulating in the community. However, in the absence of nationally licensed COVID-19 vaccines and curative treatments, researchers remained reluctant to recommence recruitment, owing to the on-going risk of a possible COVID-19 outbreak, with attendant risks to older people. They, and their community partners, were unclear about how to determine when recommencement of the trial would be justified.

Questions

1. What ethical considerations and other factors should determine when, and whether, to recommence this trial?
2. Should a higher level of caution regarding the risks of restarting CST apply if it's being provided as part of a research study rather than as routine care? Why? Are characteristics of the participant group (such as a dementia diagnosis) relevant to this question?
3. How should the potential benefit of participation in this trial figure in reasoning about when recommencement is justified?
4. Should funding considerations influence decisions about whether and when to recommence this study? Why?

Case 5.3: Suspending Participation in Research

This case study was written by members of the case study author group.

Keywords Research design and adaptation; Safety and participant protection; Risk/benefit analysis; Researcher roles and responsibilities; Access to experimental treatments; Vulnerability and inclusion; Digital and remote healthcare and research; Non COVID-19 research

During the COVID-19 pandemic the population of a Latin American country was vulnerable, owing to a very high prevalence of cardio-metabolic illness. Such illnesses are a major risk factor for a bad prognosis following infection with the SARS-CoV-2 virus because of their effect on the immune system and the chronic inflammatory state they promote (Carter et al. 2020; Jayawardena et al. 2020).

Research centres in the country are, in general, responsible for the health of their research participants. Some research centres also provide clinical care, and thus have responsibilities to their patients as well. It is important for them to uphold institutional directives and guidelines for infection control in order to minimize risks to patients and research participants. In the pandemic, novel ways of conducting research have had to be developed to decrease potential exposure to COVID-19. Some studies introduced home delivery systems for medicines and equipment so that participants were not unnecessarily exposed to infection at a research centre or clinic. However, the storage and handling of research products is a delicate process, which must follow adequate procedures and be appropriately documented.

During the pandemic a Phase III pharmacological study of moderate rheumatoid arthritis was conducted. To reduce the risk of infection, scheduled visits to the research centre by participants who reported that they had been infected with COVID-19 were replaced by telephone calls. Participants who did not report COVID-19 infection continued with routine face-to-face visits. A participant in the study experienced a 20% improvement in swelling and pain in his joints (when compared to baseline measurements at his first visit). At approximately week 84 of the study the participant was diagnosed with COVID-19. At this point his participation in the study was halted, as the medication being trialled had immunosuppressive activity.

In the context of evolving scientific understanding of COVID-19, the researchers needed to consider preventive measures that might need to be implemented to avoid furthering risks to participants. In this case questions arose about whether participants in a study of a treatment with immunosuppressive effects should resume research participation after their COVID-19 infection had resolved, and if so, when. The potential effects of the medication being trialled on both the severity of the rheumatoid arthritis, and the length and severity of COVID-19 symptoms were considered.

Questions

1. What ethical considerations should be taken into account when making decisions about the potential suspension and recommencement of research participation following a participant's COVID-19 diagnosis? What role, if any, should participants' views and preferences have in such decisions?
2. How should research centres prioritize research addressing chronic health needs in a pandemic? What specific considerations may arise when the proposed research involves a medication which has the potential to address chronic health problems but has immunosuppressive activity?
3. What responsibilities should researchers have to participants who are unable to continue receiving study interventions following a diagnosis of COVID-19?
4. What ethical issues should be considered when deciding if research studies should implement home delivery systems for research equipment and medication to minimize the number of site visits needed during a pandemic?

References

Carter, S.J., M.N. Baranauskas, and A.D. Fly. 2020. Considerations for obesity, vitamin D, and physical activity amid the COVID-19 pandemic. *Obesity* 28: 1176–1177. https://doi.org/10.1002/oby.22838.

Jayawardena, R., P. Sooriyaarachchi, M. Chourdakis, C. Jeewandara, and P. Ranasinghe. 2020. Enhancing immunity in viral infections, with special emphasis on COVID-19: A review. *Diabetes & Metabolic Syndrome: Clinical Research & Reviews* 14: 367–382. https://doi.org/10.1016/j.dsx.2020.04.015.

Case 5.4: Ethics and Adaptive Trials in the COVID-19 Pandemic

This case study was written by members of the case study author group.

Keywords Research design and adaptation; Vulnerability and inclusion; Consent; Ethical review; Researcher roles and responsibilities; Research priority setting; Risk/ benefit analysis; Research publication ethics; Treatment repurposing; Multi-centre research; Pre-prints

A broad range of potential treatments have been proposed for COVID-19. Conducting timely and rigorous research to evaluate the effectiveness and safety of possible treatments is key to informing effective public health responses to COVID-19. The Randomised Evaluation of COVID-19 Therapy (RECOVERY) study is an adaptive randomized controlled trial, initially established in the UK to evaluate which treatments may be more effective than the usual standard of care patients receive when admitted to National Health Service (NHS) hospitals with COVID-19 (University of Oxford 2021a).

Adaptive trials are so called because the protocol pre-specifies that certain elements of the trial may be adapted during the trial. In the RECOVERY trial, the key adaptive element is that new study arms are added when there is sufficient reason to believe an intervention may offer a benefit but when there is also uncertainty. Study arms are closed once sufficient data have been collected to establish whether the intervention is associated with a benefit, or lack of benefit or even harm, for participants. Large-scale adaptive trials can be powerful and give results to inform policy more efficiently than traditional clinical research designs (Pallmann et al. 2018). During the first 11 months of RECOVERY, for example, over 38,000 participants were recruited at over 170 sites. By February 2021, the study had published preliminary or complete findings about the benefit or lack of benefit associated with dexamethasone, tocilizumab, hydroxychloroquine, lopinavir- ritonavir, convalescent plasma and azithromycin in the treatment of hospitalized COVID-19 patients (University of Oxford 2021b). On-going study arms were assessing the benefits of monoclonal antibodies, aspirin, colchicine, baricitinib and dimethyl fumarate.

Review

Adaptive trials can be challenging to review for ethics committees and regulatory bodies, owing to their complexity and on-going evolution; by January 2021, RECOVERY was on version 13 of the protocol. During public health emergencies, such challenges can be exacerbated by the need for effective multi-site review within an expedited timeframe. In the UK, the RECOVERY protocol and regular protocol amendments have received expedited review by a single ethics committee, which provides national approval for the research to be conducted at all participating NHS sites. National regulatory review of RECOVERY is undertaken using routine review processes within an accelerated timeframe. To support these expedited regulatory and ethical review processes, RECOVERY's principal investigators liaise with reviewers to provide updates about proposed protocol amendments and the timelines within which they will be submitted.

Public Health Responses

Within public health emergencies, the need to conduct research to address public health priorities must be evaluated in conjunction with the importance of ensuring that such research does not adversely impact pandemic response efforts (Nuffield Council on Bioethics 2020; WHO 2020). A key priority in the design and implementation of RECOVERY has been to minimize the impact of such a large-scale study on the provision of clinical care within NHS hospitals. Practical measures to achieve this include the development of a short and simple online case report form for data collection, substantial resources to support site staff, and the use of linkage to routine health data to collect information on patient outcomes (University of Oxford 2021c).

Inclusion

Current standards in research ethics highlight the need for fair and inclusive approaches to the selection of research participants, recognizing the importance of generating relevant evidence to inform approaches to addressing their health needs, as well as ensuring appropriate protections are in place (CIOMS 2016). Within the context of public health emergencies, the importance of inclusive approaches has been highlighted (Nuffield Council on Bioethics 2020), including, for example, calls to ensure that pregnant women are appropriately included in COVID-19 treatment trials (Taylor et al. 2021). All patients admitted to NHS hospitals in the UK with suspected or confirmed COVID-19 are potentially eligible to participate in RECOVERY. Potential participants will be excluded from the trial if their attending clinician believes that their medical history puts them at significant risk if they participate, or if one of the active treatment arms in the trial is considered specifically indicated or contra-indicated for the patient. Age-related exclusion criteria are associated with some treatment arms, and in trial sites outside the UK, but across the study as a whole, participant ages range from less than 1 year old to 103. Pregnant and breastfeeding women are excluded from treatment arms incorporating contra-indicated interventions but are eligible for other arms (https://www.recoverytrial. net/for-site-staff/site-teams).

Consent

Seeking consent to adaptive trials raises a number of practical ethical issues associated with the complexity of the research, on-going amendments to study arms, and evolving evidence about the risks and benefits of specific interventions (Global Forum on Bioethics in Research 2017). Within the RECOVERY trial, recruitment processes additionally need to be responsive to potential limitations on participants' capacity to consent to research, including in children, very elderly patients and patients with severe disease. Participant information sheets for adults, children, parents and guardians have been designed to be clear and concise – no more than three pages long. Information for patients is also available on the RECOVERY website, including videos describing the interventions being evaluated (https://www.recoverytrial.net/study-faq). Resources for trial sites include training and standard operating procedures for the recruitment of competent and incompetent patients (https://www.recoverytrial.net/for-site-staff/training/background-and-informed-consent).

Selecting Study Interventions

During the pandemic, a wide range of therapies have been proposed to treat COVID-19, with varying quality of supporting evidence. Given the scale and pace of global COVID-19 research, the evidence base about specific treatments has frequently evolved rapidly. Within this complex context, RECOVERY investigators make decisions about which treatments to trial, and review recommendations from the study's independent data-monitoring committee about appropriate points to halt recruitment into specific trial arms (https://www.recoverytrial.net/for-site-staff/site-set-up-1/data-monitoring-committee-correspondence). Choices of which treatments to trial are taken by RECOVERY's principal investigators and the UK Chief Medical Officer, and have been informed by WHO priorities, reviews from the UK's New and Emerging Respiratory Virus Threats Advisory Group and the UK COVID-19 Therapeutics Advisory Panel. Key considerations include existing evidence about the safety and efficacy of the intervention, whether it is available in sufficient quantities to evaluate in the trial, and whether, if shown to be successful, treatment could be rapidly scaled up (Wise and Coombes 2020).

Reporting Research Results

The pre-publication and publication of research results can have a rapid, substantial and multinational effect on research and public health responses in an epidemic (Hofmann 2020). Adaptive trials seek to produce findings more rapidly than traditional clinical trials, and when there is sufficient evidence to justify halting recruitment into a specific treatment arm, questions arise about when and how preliminary findings should be disseminated. Preliminary findings from RECOVERY are typically reported in press releases (https://www.recoverytrial.net/news/) and simultaneous pre-prints, with study results subsequently being published in peer-reviewed journals (https://www.recoverytrial.net/results/).

Questions

1. What specific ethical issues arise when conducting consent processes for an adaptive trial in a pandemic, and how should they be addressed?
2. Whose perspectives should inform decisions about which potential treatments should be assessed in the RECOVERY trial, and what ethical considerations should inform that decision-making?
3. How should researchers and health-care providers respond ethically to press releases about preliminary research results during a pandemic? Why?
4. What ethical arguments could be made for and against conducting expedited national ethical reviews of RECOVERY protocols and amendments (rather than review at each trial site)?

References

CIOMS. 2016. International ethical guidelines for health-related research involving humans. Geneva: Council for International Organizations of Medical Sciences.

Global Forum on Bioethics in Research. 2017. Ethics of alternative clinical trial designs and methods in LMIC research. https://www.gfbr.global/wp-content/uploads/2018/12/GFBR-2017-meeting-report-FINAL.pdf.

Hofmann, B. 2020. The first casualty of an epidemic is evidence. *Journal of Evaluation in Clinical Practice* 26(5): 1344–1346.

Nuffield Council on Bioethics. 2020. *Research in global health emergencies*. London: Nuffield Council on Bioethics.

Pallmann, P., et al. 2018. Adaptive designs in clinical trials: Why use them, and how to run and report them. *BMC Medicine* 16(1): 29.

Taylor, M.M., et al. 2021. Inclusion of pregnant women in COVID-19 treatment trials: A review and global call to action. *The Lancet Global Health* 9(3): e366–e371.

University of Oxford. 2021a. RECOVERY: Randomised evaluation of COVID-19 therapy. https://www.recoverytrial.net/.

University of Oxford. 2021b. RECOVERY: Randomised evaluation of COVID-19 therapy. https://www.recoverytrial.net/results.

University of Oxford. 2021c. RECOVERY: Randomised evaluation of COVID-19 therapy. https://www.recoverytrial.net/for-site-staff.

WHO. 2020. *Ethical standards for research during public health emergencies: Distilling existing guidance to support COVID-19 R&D*. Geneva: World Health Organization.

Wise, J., and R. Coombes. 2020. Covid-19: The inside story of the RECOVERY trial. *British Medical Journal* 370: m2670.

Case 5.5: The Impact of New Scientific Evidence on On-going COVID-19 Studies

This case study was written by members of the case study author group.

Keywords Research design and adaptation; Pausing and halting research; Ethics committee remits and responsibilities; Ethical review; Researcher roles and responsibilities; Social and scientific value; Treatment repurposing

A national research ethics committee (NREC) reviewed and approved three local research initiatives testing the efficacy of hydroxychloroquine for COVID-19: one for patients with mild to moderate COVID-19 (*Trial A*), another for patients with critical and severe COVID-19 (*Trial B*), and a third one that tested hydroxychloroquine as prophylaxis (*Trial C*).

After the approval of these trials, the preliminary results of a randomized, controlled, open-label adaptive trial demonstrated that there was no clinical benefit from administering hydroxychloroquine to hospitalized people with COVID-19 (University of Oxford 2021) (see also Case 5.4). The enrolment of participants to this arm of the trial was subsequently stopped. After the release of these data, another global clinical trial also suspended its hydroxychloroquine arm in order to analyse its interim results, and finally cancelled the arm because continuation was considered futile (WHO n.d.).

Some members of the NREC heard through news media about the measures adopted in these trials in response to the new scientific evidence and shared this information immediately with the committee. The NREC scheduled an extraordinary meeting on the following day to discuss the implications of this information. The committee considered that these events might have an impact on the ethical acceptability of the three trials it had approved (PAHO 2020). However, some members were not sure how the NREC should respond, given that it had not had access to the details of the new evidence. The committee finally decided that as it is a researcher's responsibility to report such information to the NREC, they should wait to receive the researchers' reports. They also agreed that in the midst of the pandemic, their priority was to review COVID-19 protocols recently submitted for ethics approval rather than to ask for more reports from previously approved studies.

The updates that the NREC received from the researchers of the trials testing hydroxychloroquine varied, as indicated below:

- *Trial A*: After 3 days, the principal investigator informed the NREC that they had suspended the trial in order to evaluate the impact of the new data on the efficacy and safety of hydroxychloroquine for their participants, who had mild to moderate COVID-19. Given that the new evidence was specifically for hospitalized COVID-19 patients, the research team was not sure if they were justified in extrapolating these results to the participants of their trial.
- *Trial B*: After 2 weeks, the principal investigator sent the periodic progress report according to the deadline that had been previously established by the NREC. The

report stated that 23 participants were enrolled, and that the trial was being conducted in accordance with the protocol. No mention was made of the recent reports from other trials.

- *Trial C*: Three weeks after reports emerged about the suspension of the other hydroxychloroquine trials, the NREC had not received any information from the research team.

In response, the NREC took the following action. For *Trial A*, the committee asked the principal investigator to communicate to the participants the reasons for the suspension of the trial, and to inform the NREC of their final decision regarding the continuation of the trial as soon as they had finished analysing the evidence. For *Trials B and C*, the NREC asked the principal investigators to justify the continuation of the study on the basis of the newly available evidence. In the case of *Trial C*, the principal investigator replied that he had not provided any report because the new evidence did not affect the rationale for testing hydroxychloroquine as prophylaxis for COVID-19.

Questions

1. What are the implications of the rapid production of new evidence during a pandemic for the ethics oversight of COVID-19 research?
2. What are the responsibilities of COVID-19 researchers with respect to emerging scientific evidence that could affect the justification for conducting their research and/or how it is conducted?
3. Should research ethics committees adopt special operating procedures for the oversight of ongoing COVID-19 research? If so, what should these procedures include and why?
4. How should the ethical analysis of on-going protocols be conducted in light of new evidence, in order to ensure their continuing ethical acceptability? What questions should guide this assessment?

References

PAHO. 2020. Guidance for ethics oversight of COVID-19 research in response to emerging evidence. Washington: Pan American Health Organization. https://iris.paho.org/handle/10 665.2/53021.

University of Oxford. 2021. RECOVERY: Randomised evaluation of COVID-19 therapy. https://www.recoverytrial.net/.

WHO. n.d. "Solidarity" clinical trial for COVID-19 treatments. World Health Organization. https://www.who.int/emergencies/diseases/novel-coronavirus-2019/global-research-on-novel-corona virus-2019-ncov/solidarity-clinical-trial-for-covid-19-treatments.

References

Adebamowo, C., O. Bah-Sow, F. Binka, R. Bruzzone, A. Caplan, J.-F. Delfraissy, D. Heymann, P. Horby, P. Kaleebu, J.-J. Muyembe Tamfum, P. Olliaro, P. Piot, A. Tejan-Cole, O. Tomori, A. Toure, E. Torreele, and J. Whitehead. 2014. Randomised controlled trials for Ebola: Practical and ethical issues. *The Lancet* 384(9952): 1423–1424.

Cartwright, N. 2007. Are RCTs the gold standard? *BioSocieties* 2: 11–20.

Ezeome, E., and C. Simon. 2010. Ethical problems in conducting research in acute epidemics: The Pfizer meningitis study in Nigeria as an illustration. *Developing World Bioethics* 10(1): 1–10.

Hemming, K., T. Haines, P. Chilton, A. Girling, and R. Lilford. 2014. The stepped wedge cluster randomised trial: Rationale, design, analysis, and reporting. *British Medical Journal* 350(h391).

Joffe, S. 2014. Evaluating novel therapies during the Ebola epidemic. *Journal of the American Medical Association* 312(13): 1299–1300.

Kass, N. 2014. Ebola, ethics, and public health: What next? *Annals of Internal Medicine* 161(10): 744–745.

London, A.J., and J. Kimmelman. 2020. Against pandemic research exceptionalism. *Science* 368(6490): 476–477.

Mahajan, R., and K. Gupta. 2010. Adaptive design clinical trials: Methodology, challenges and prospect. *Indian Journal of Pharmacology* 42(4): 201–207.

Modlin, C., J. Sugarman, G. Chongwe, N. Kass, W. Nazziwa, J. Tegli, P. Shrestha, and J. Ali. 2023. Towards achieving transnational research partnership equity: Lessons from implementing adaptive platform trials in low- and middle-income countries. *Wellcome Open Research* 8: 120. https://doi.org/10.12688/wellcomeopenres.18915.1.

Nuffield Council on Bioethics. 2020. *Research in global health emergencies*. https://nuffieldbioethics.org/publications/research-in-global-health-emergencies.

PAHO. 2020. *Guidance for ethics oversight of Covid-19 research in response to emerging evidence*. Pan American Health Organization. https://iris.paho.org/handle/10665.2/53021

Pallmann, P., A.W. Bedding, B. Choodari-Oskooei, M. Dimairo, L. Flight, L.V. Hampson, J. Holmes, A.P. Mander, L. Odondi, M.R. Sydes, S.S. Villar, J.M. Wason, C.J. Weir, G.M. Wheeler, C. Yap, and T. Jaki. 2018. Adaptive designs in clinical trials: Why use them, and how to run and report them. *BMC Medicine* 16(29). https://doi.org/10.1186/s12916-018-1017-7.

Pullman, D., and X. Wang. 2001. Adaptive designs, informed consent, and the ethics of research. *Controlled Clinical Trials* 22(3): 203–210.

Singh, J.A. 2023. Adaptive clinical trials in public health emergency contexts: ethics considerations. *Wellcome Open Research* 8: 130. https://doi.org/10.12688/wellcomeopenres.19057.1.

Tangwa, G. 2014. Ebola epidemic: The WHO gets it right, then wrong. https://www.gobata.com/2014/10/ebola-epidemic-the-who-gets-it-right-then-wrong.html.

WHO. 2014. Ethical considerations for use of unregistered interventions for Ebola viral disease. World Health Organization. https://www.who.int/news/item/12-08-2014-ethical-considerations-for-use-of-unregistered-interventions-for-ebola-virus-disease-(evd).

———. 2022. Emergency use of unproven clinical interventions outside clinical trials: ethical considerations. World Health Organization. https://www.who.int/publications-detail-redirect/9789240041745.

Chapter 6
Ethics Review Challenges

Sarah Carracedo, Ana Palmero, and Carla Saenz

Abstract In the context of a public health emergency it is imperative to conduct research studies that will produce evidence rapidly while upholding ethical standards. The Ebola and Zika outbreaks highlighted the importance of devising agile processes for ethics review in emergencies, and international research ethics guidelines stress the duty to depart from standard processes for ethics review in emergency circumstances. However, before the COVID-19 pandemic it was not entirely clear what emergency procedures should look like. An additional challenge is that while the same substantive ethical standards apply in emergency and non-emergency settings, deciding what these standards entail in the specific circumstances of a pandemic may be difficult. During the COVID-19 pandemic, challenges included identifying thresholds of social and scientific value, along with duties towards research participants, given the absence of therapeutic options; assessing continually changing risk–benefit profiles of studies, given rapidly emerging new evidence; developing appropriate informed consent processes, given lockdown scenarios; and even addressing the ethics of offering unproven interventions outside research settings. Additional issues raised during epidemics include devising feasible and meaningful community engagement strategies, mechanisms to ensure fairness in the distribution of the benefits that may result from research, and equitable and effective data-sharing plans that will inform pandemic response. Learning from these procedural and substantive challenges encountered in the ethics review of COVID-19 research is important for enhancing ethics preparedness for future emergencies. It can also potentially contribute to improving the ethics review of research in

S. Carracedo · C. Saenz (✉)
Regional Program on Bioethics, Pan American Health Organization (PAHO),
Washington, DC, WA, USA
e-mail: saenzcar@paho.org

A. Palmero
Directorate of Research for Health, Ministry of Health, Buenos Aires, Argentina

© PAHO and Editors 2024
S. Bull et al. (eds.), *Research Ethics in Epidemics and Pandemics: A Casebook*,
Public Health Ethics Analysis 8, https://doi.org/10.1007/978-3-031-41804-4_6

non-emergency circumstances. The seven cases in this chapter highlight ethical issues associated with ethics approval of multi-centre studies in pandemics, the need for careful consideration of the social and scientific value of research and challenges encountered when interventions are being transitioned from research to rollout, and issues that can arise when existing regulations and policies may limit capacities to appropriately adapt research to pandemic contexts.

Keywords COVID-19 pandemic · Research ethics · Public health emergencies · Ethics review · Research ethics committee remits and responsibilities · Risk/benefit analysis · Data protection, access and sharing · Safety and participant protection · Research priority setting · Research quality · Social and scientific value · Digital and remote healthcare and research · Emergency Use Authorisation · Regulatory review

6.1   Introduction

Research is an essential component of the public health response to an emergency. We need knowledge in order to find safe and efficacious interventions to help us understand, prevent, diagnose and treat emerging diseases, and overall to guide the public health response. It is imperative to conduct research studies that will produce evidence rapidly while upholding ethical standards.

However, conducting research in emergency settings is challenging. Some challenges pertain to the process of ensuring rigorous yet rapid ethics review during emergencies. Ethics review by an independent research ethics committee (REC) aims to ensure the ethical conduct of research with human participants. While obtaining ethics approval before the start of a study is a requirement that must not be bypassed during health emergencies – and doing so is tantamount to conducting research unethically – RECs should streamline their processes in order to conduct ethics review in a timely manner during a pandemic. Previous public health emergencies of international concern (PHEIC), like the Ebola and Zika outbreaks, highlighted the importance of devising agile processes for ethics review in emergencies, and the 2016 guidelines published by the Council for International Organizations of Medical Sciences (CIOMS) stress the duty to depart from standard mechanisms for ethics review in emergency circumstances (CIOMS 2016). However, before the COVID-19 pandemic it was not entirely clear what emergency procedures should look like.

A different set of challenges pertains to the ethics analysis conducted by RECs when reviewing health emergency research protocols. While the same substantive ethical standards apply in emergency and non-emergency settings, as stressed in the CIOMS guidelines and other guidance documents (CIOMS 2016; PAHO 2016, 2020b, c, f; WHO 2016, 2020a), deciding what these standards entail in the specific circumstances of a pandemic may be difficult. During the COVID-19 pandemic, challenges included identifying thresholds of social and scientific value, along with duties towards research participants, given the absence of therapeutic options; assessing continually changing risk–benefit profiles of studies, given rapidly emerging new evidence; developing appropriate informed consent processes, given lockdown scenarios; and even addressing the ethics of offering unproven interventions outside research settings.

Learning from these procedural and substantive challenges encountered in the ethics review of COVID-19 research is important for enhancing ethics preparedness for future emergencies. It can also potentially contribute to improving the ethics review of research in non-emergency circumstances.

6.2 Challenges in the Ethics Review Processes

During the COVID-19 pandemic, several guidance documents about ethical COVID-19-related research involving human subjects were issued to guide national authorities and RECs during the health emergency (PAHO 2020b, c, f; WHO 2020a, b). The first and most important task identified in these documents was for countries to establish a strategy for the organization of ethics review and oversight that was best suited to their context. This was to be undertaken with the aim of avoiding duplication of effort, preventing RECs from becoming overwhelmed, and developing mechanisms for coordination and communication between the relevant research stakeholders. For example, the relevant authorities could decide to create an ad hoc committee tasked with the conduct of ethics review and the oversight of COVID-19 research, or designate an existing REC of a national entity to be responsible for the review of these protocols. Another option could be to task one or more institutional RECs with the review of (certain types of) COVID-19 research (PAHO 2020c, 2022). Indeed, several countries rapidly implemented these strategies and, overall, international guidance to accelerate the ethics review of COVID-19 research (ICMR 2020; Palmero et al. 2021; PAHO 2022).

It is important to consider that any adopted strategy will need to be supplemented by "emergency mode" operating procedures in order to ensure that the RECs tasked with the review of COVID-19 protocols conduct a rapid yet rigorous review. Such procedures include tight deadlines for reviews, virtual meetings, electronic submission of research proposals, the inclusion of additional members and subject experts, and mechanisms for coordination and communication between RECs, investigators and authorities, among others (PAHO 2020c, f, 2022; WHO 2020b).

The difficulties that can arise in the absence of planning and of establishing a national strategy to organize and streamline ethics review and oversight processes are illustrated in Cases 6.1 and 6.2. As shown by both cases, this is of special relevance in the case of multi-centre studies because the involvement of several RECs without rapid and flexible operating procedures or clear mechanisms of coordination and communication can result in practical obstacles that duplicate efforts, waste time and resources and, ultimately, result in missing valuable research opportunities.

6.3 Challenges in the Ethics Analysis of Health Emergency Research

Ethics review of emergency research proposals must adhere to existing national and international ethical standards. However, RECs face several challenges when conducting their ethics analyses because the emergency context may seem to justify flexibility where ethical standards are concerned, when in fact it should highlight the importance of increased diligence in review *processes*. Certainly, applying ethical standards to unusual and rapidly changing contexts, like the COVID-19 pandemic, can be challenging and requires strong capacities for ethics analysis.

Cases 6.3–6.6 illustrate some of the challenges faced by RECs when assessing the anticipated social and scientific value of research. In addition, the cases invite reflection on the challenges to the ethical conduct of research in cases where interventions have been proven safe and efficacious elsewhere and are being transitioned from research to rollout, and on the role of RECs regarding the use, outside of research settings, of interventions that have not been proven. Case 6.7 raises issues about the informed consent process during the COVID-19 pandemic, which is another component of ethics review.

6.3.1 Social and Scientific Value of Research

Despite regulatory, logistical and practical difficulties in emergency situations, research with human participants must have social and scientific value. Lack of purpose and scientific rigour may not just waste resources and effort but it is ethically problematic because participants are being exposed to risks without the prospect of future benefits, such as valuable and valid knowledge, being produced. The urgency with which knowledge is needed in emergency situations does not alter moral duties to conduct research that adheres to scientific standards. Such urgency should not be construed as permission to conduct research that is not scientifically sound or that is ethically questionable for any other reason (London and Kimmelman 2020).

During the COVID-19 pandemic, we have seen a vast amount of research conducted in an unprecedentedly short time. Yet many of the studies have small samples, unnecessarily replicate hypotheses, do not have adequate comparators, and in general, have methodological flaws that make them unlikely to produce robust evidence or even incapable of producing it (London and Kimmelman 2020; Carracedo et al. 2020). Moreover, the choice of some interventions under study does not seem justified by prior knowledge, and even interventions already known to be harmful have been studied (London 2021; PAHO 2020e; Herper and Riglin 2020) (see Chap. 3).

Cases 6.3–6.5 illustrate the need for careful consideration of the social value and scientific merits of protocols as part of the ethics review. Challenges pertaining to the soundness of the scientific justification for the research, and the acceptability of small and repetitive clinical trials that do not seem capable of producing robust evidence in a timely manner, as well as the qualifications of the research team to conduct emergency research, highlight the importance of the role of RECs in ensuring well-designed and high-quality research that can generate valuable knowledge during a health emergency.

Another crucial aspect to take into account is the fact that the social and scientific value of research is not static, and that the ethical acceptability of approved protocols can vary while the study is underway. This is especially relevant in emergency settings characterized by a rapid production of scientific evidence, as discussed in Chap. 5. In this sense, the oversight of on-going emergency research may need to be more frequent. That means, on the one hand, that researchers have a responsibility to constantly evaluate the justification for their research on the basis of the most up-to-date evidence and to make decisions regarding the conduct of their studies accordingly. And on the other hand, RECs should oversee the conduct of the study until its completion, in light of the emerging scientific evidence, in order to take appropriate measures to guarantee its continued adherence to ethical standards (PAHO 2020d, 2022).

6.3.2 Consent Processes

Obtaining informed consent from participants is necessary for all research involving human subjects conducted during a health emergency, unless an REC approves a waiver of this requirement on the basis of particular criteria, such as those established by the CIOMS guidance (CIOMS 2016).[1] In practice, several obstacles caused by the circumstances of the health emergency (e.g. isolation of patients, lockdowns) may preclude the ordinary process of obtaining informed consent and thus pose a need to consider alternative ways of doing so. Indeed, during the COVID-19 pandemic, alternative procedures were proposed (PAHO 2020c, f) and established by many national authorities and regulatory agencies (ICMR 2020; Palmero et al. 2021). For example, the use of electronic informed consent forms or other electronic means (e.g. telemedicine, phone calls, video calls, photographs, etc.) to facilitate consent processes and to contact family members or legal representatives for support or proxy consent has become increasingly widely accepted (see Chap. 9).

However, in many cases these alternative processes have not been deemed compliant with national regulatory frameworks that stipulate requirements for consent processes (e.g. an in-person process) and the documentation of research participants' decisions (e.g. in a hard copy). Therefore, it is necessary to think about how regulatory requirements for consent processes should be addressed in emergency

---

[1] According to Guideline 10, a waiver of informed consent to research may be approved by an REC when the research proposal (a) would not be feasible or practicable to carry out without the waiver, (b) has important social value, *and* (c) poses no more than minimal risks to participants.

situations, which were presumably not taken into account when these requirements were devised. Consequently, RECs may need to be prepared to reassess the conduct of informed consent processes and suggest alternative methods. This requires RECs to understand and balance the differences between their responsibility to ensure respect for potential participants' autonomy and the need to comply with legal requirements, some of which may even expose participants and researchers to higher risks (e.g. the possible spread of the disease through paper forms or in-person encounters). Case 6.7 raises this issue and invites us to reflect on what it is right to do when existing regulations and institutional policies may come into conflict with ethics.

6.3.3   Research in the Transition from Research to Rollout

The magnitude of research that has been conducted in response to COVID-19 has led to the rapid production of evidence on the safety and efficacy of interventions, and subsequent approvals of these interventions by different national regulatory authorities (NRAs) around the world. NRAs have often relied on emergency use authorizations (EUAs) in order to make therapeutic and preventive interventions available as soon as possible during the pandemic.

In this scenario, the conduct of research has encountered additional challenges. Some challenges pertain to the justifications for conducting research on interventions that have been deemed safe and efficacious in other jurisdictions and are being offered to the population there. Several reasons to justify research on these interventions can be conceived, e.g. to obtain more specific knowledge, which initial studies were not designed to obtain, including their performance in response to new SARS-CoV-2 variants. Arguably, an EUA highlights the need to continue conducting research, for instance to collect follow-up data for a longer period of time. Additionally, the conduct of local research in order to authorize interventions may be a national regulatory requirement, even if those interventions have been proven safe and efficacious elsewhere and authorized by other NRAs. In any case, the justification for research in these circumstances should be clearly laid out and the proposals carefully assessed, to ensure the ethical conduct of the study.

Case 6.6 illustrates the complex set of ethical challenges that can be encountered when research is conducted while the interventions under study are being rolled out as part of the delivery of care. These challenges range from the justification for these studies, to the duties towards research participants and their communities. In the context of COVID-19, there has been a global need for efficacious interventions for prevention and treatment, but these interventions have not been available everywhere at the same time. Moreover, the availability of these interventions as part of the delivery of care can hamper the capacity to conduct research (such as when a study entails randomization) even if such research is associated with significant social and scientific value and could otherwise be conducted ethically. RECs must conduct a careful assessment of these studies and may have to grapple with the ethical justification of regulatory requirements (such as bridging trials) in emergency situations. Furthermore, as Case 6.6 points out, there is a need to consider regulatory procedures that can catalyse rigorous and rapid research during emergencies, as has been done with ethics review and oversight processes.

6.3.4 Use of Unproven Intervention Outside Research Settings

Another challenge faced during public health emergencies is the exceptional use of unproven interventions outside research settings (see Chap. 4). In normal circumstances, interventions are tested in research settings to prove their safety and efficacy. In these settings, as participants are receiving interventions whose safety and/or efficacy is being evaluated, safeguards are necessary (e.g. ethics review by a REC). Once an intervention has been proven safe and efficacious, it can be offered to persons outside research settings, i.e. as a public health intervention without these safeguards.

However, during health emergencies, unproven interventions are sometimes offered outside research settings as an alternative that *may* benefit patients. This raises obvious concerns: the safeguards intrinsic to research settings are not available outside such settings yet people are receiving interventions whose safety and efficacy are not known. The use of unproven interventions in these exceptional contexts *can*, however, be ethically justified if it adheres to ethical and scientific standards aimed at protecting patients and affected populations from the risks involved in such interventions.

During the Ebola outbreak, the World Health Organization (WHO) developed a framework for the ethical use of unproven interventions outside research settings, involving collecting data on patient outcomes to contribute to the generation of new knowledge (WHO 2016). The framework was referred to as "monitored emergency use of unregistered and experimental interventions" (MEURI). It was later refined by the Pan American Health Organization (PAHO) to provide actionable ethical guidance for the COVID-19 pandemic (PAHO 2020a, 2022). To determine if it is ethically acceptable to offer unproven interventions outside research situations, PAHO organized the existing WHO criteria into four categories: the justification for the intervention, its ethical and regulatory oversight, an informed consent process and its contribution to the generation of evidence.

Yet the ethics framework for the emergency use of unproven interventions outside research was not widely known before the COVID-19 pandemic. At the inception of the pandemic, many RECs and relevant health authorities were not familiar with the criteria for determining whether it was ethical to offer unproven interventions outside research and what their role should be (e.g. whether RECs should review such proposals). This may partially explain why we have witnessed the use of several interventions, such as hydroxychloroquine, convalescent plasma, ivermectin, and even chlorine dioxide, outside research contexts without adherence to relevant ethics guidance (see Case 3.2 in Chap. 3). Case 6.5 illustrates the challenges faced by RECs when a proposal for use of an unproven intervention outside a research context is submitted for ethics review. It highlights the need for ethics guidance addressing this exceptional situation, which may arise during emergencies, including the appropriate roles of RECs.

6.4   Conclusions

A thorough assessment of the lessons learned during the COVID-19 pandemic can play an important role in strengthening ethics preparedness for future emergencies around the world. As illustrated in this chapter, ethics review challenges require more than implementing procedures to make ethics reviews faster and strategies to avoid the duplication of effort. The quality of the ethics analysis and the ability to adapt and respond to the emergency environment is also very important. The capacities of REC members may need to be strengthened so they are better prepared to conduct rigorous ethics reviews of emergency research in a timely manner.

The challenges that RECs have faced during the COVID-19 pandemic go beyond what these cases have illustrated. As discussed in other chapters in this casebook, additional issues raised in epidemics which need to be addressed to ensure that research is conducted ethically include devising feasible and meaningful community engagement strategies, mechanisms to ensure fairness in the distribution of the benefits that may result from research, and equitable and effective data-sharing plans that will inform pandemic response. Addressing these issues appropriately is essential for sustaining trust in research and, furthermore, in the interventions that result from research, which COVID-19 has revealed as crucial for ending the pandemic. Finally, we should critically evaluate which practices that were implemented in response to COVID-19 and have shown success (e.g. virtual meetings of RECs, electronic informed consent processes) and should be adopted into non-emergency ethics review and oversight processes. Lessons learned from this pandemic go beyond the ethics of emergency research and may help us to improve ethics review of research in the future.

Case 6.1: Ethics Approval of a Multi-centre Study: To Expedite or Not?

This case study was written by members of the case study author group.

Keywords Ethics review; Research ethics committee remits and responsibilities; Risk/benefit analysis; Data protection, access and sharing; Safety and participant protection; Research priority setting; Multi-centre research; Treatment repurposing

While there were a number of licensed COVID-19 vaccines in use globally during 2021, the high cost and short supply meant that they were not expected to be readily available to the majority of the population in an Asian country until the end of 2021. Before that there remained an urgent need to find effective ways to continue treating patients.

In December 2020, the number of COVID-19 infections in the Asian country began to surge despite strict lockdown measures. Dr X. was a world-renowned medical researcher from an Asian country with extensive experience and knowledge of infectious diseases. She was appointed as a member of the COVID-19 task force in her institution and participated in a number of international COVID-19 initiatives. Some of the patients in Dr X.'s care developed cytokine storm syndrome, a common side-effect of severe COVID-19 and one which is potentially life-threatening. At the peak of the outbreak, Dr. X. and her team treated these patients with an IL-6 (interleukin 6) receptor blocker and saw tremendous improvements. Dr. X. began to design a randomized control trial to test whether this was an effective treatment for cytokine storm syndrome. The study participants would be divided into two groups, one receiving the IL-6 blocker and the control group receiving high doses of steroids as standard care. In order to increase the number of participants for better statistical analysis, Dr X. sought to expand her trial to include two other tertiary-level hospitals in the country which were COVID-19 centres.

Given the multi-site nature of the trial, Dr X. was required to seek ethical approval from each institution's ethics review committee as well as from the national ethics review committee. Dr X. and her team were hoping that expedited pathways for COVID-19 trials would enable them to get approval within 2 weeks so that they could begin the clinical trial as soon as possible. Dr X. strongly believed that the findings would support her hypothesis that the IL-6 blocker would revolutionize the treatment of cytokine storm syndrome and save many lives. The ethics review committee at Dr X.'s institution provided ethical approval within 1 week. However, the two other ethics review committees only responded 3 weeks after submission and the national ethics review committee responded 4 weeks after submission. All three committees asked for more information regarding particular issues. Committee A was concerned about the risk of using steroids as a standard of care, noting that their institution might not be able to provide them because they were short of resources and personnel. Committee B raised a concern about the possible risks to patients and requested details about compensation to participants for adverse events. This committee was concerned that the families of the participants should be compensated for any medical complication due to the novelty of the disease, as well as research-related harms. Committee C highlighted data privacy issues resulting from collation of personal information about patients recruited from the three participating COVID-19 centres.

The feedback from the ethics review committees about these concerns was reflected in their requests for different amendments to the protocol. By the time these reviews came through, the peak of infections had passed and Dr. X. was seeing far fewer COVID-19 patients. Frustrated with the process, she decided to abandon the multi-centre trial and continue the treatment modalities as a clinical practice in her own institution.

Questions
1. Do you think that the requests from the ethics committees in this case were reasonable? Why?
2. What are the challenges of reviewing a multi-centre randomized control trial in a national health emergency?
3. During a global emergency, should there be a specific process to expedite ethics reviews of research proposals involving potential treatments? What timeframes would be reasonable? Which actors should be involved? Why?
4. In a national health emergency, are multiple ethics reviews of research studies a necessary step or is there another appropriate way of seeking ethical approval for multi-centre study? Why?

Case 6.2: Ethics Review of Multi-centre Trials: Challenges and Unforeseen Issues

This case study was written by members of the case study author group.

Keywords Ethics review; Research ethics committee remits and responsibilities; Safety and participant protection; Multi-centre research

With few treatment options available to manage COVID-19 during the early stages of the pandemic, the disease presented a unique set of challenges for healthcare providers globally. Many prophylactic and therapeutic trials have been undertaken across the globe to generate evidence to inform the clinical care of patients with COVID-19.

The public health agency in an Asian country decided to investigate the effectiveness of using convalescent plasma to treat moderate COVID-19 in adults, using an open-label, parallel arm, Phase II, multi-centre, randomized controlled trial. About 50 public and private hospitals across the country expressed an interest in recruiting a total of 500 patients to the study between April and July 2020. Half the participants were randomly assigned to the intervention arm to receive convalescent plasma with standard care, and half received standard care only in the control arm. The objective of the trial was to assess the reduction in progress from mild to severe COVID-19 and the reduction in all-cause mortality. Participants in the intervention arm received two 200 ml doses of convalescent plasma, transfused 24 hours apart, in addition to standard care.

The study was coordinated centrally by a research and data coordinating centre and carried out at multiple hospital sites, using a common trial protocol. Conducting the trial presented various logistical and ethical challenges, including the following: planning the trial protocol, selecting participating sites, obtaining scientific/regulatory and ethics committee approval, conducting the trial across multiple sites, managing the challenges caused by lockdown, registering on the clinical trial registry platform, reporting serious adverse events from different sites, managing authorship issues for publication of the results, and converting evidence into practice.

All the sites willing to undertake the trial sought approval from their local ethics committee, as per standard requirements in the country. In view of the emergency situation posed by the pandemic, the trial was rapidly initiated and the recruitment commenced at all the centres as soon as the relevant ethics committee approval was received. In the country, only registered ethics committees can undertake ethics review of regulatory trials of this kind. An analysis of the status of ethics committees at these 50 sites showed that many of the trial sites did not have registered ethics committees, or had ethics committees with lapsed registration where renewal had not been applied for, or ethics committees that had applied for renewal and were awaiting their registration from the licensing authority. In addition, different ethics committees issued a range of recommended changes with respect to trial processes.

Questions
1. What logistical and ethical requirements should have been considered for the selection of the sites to participate in the trial? Should these include considerations relating to site preparedness, investigator qualifications and even local ethics committee capacity, training and/or registration? Why?
2. What should be done with the research data collected from the centres where the ethics committee registration status is invalid? Should administrative criteria like registration status be waived in an emergency like the COVID-19 pandemic? Why or why not?
3. For multi-centre clinical trials in a pandemic involving an intervention such as this, what role might a single (multi-site) ethics review play in ensuring the safety and well-being of participants? What should the division of responsibilities between local ethics committees and the multi-site common ethics committee be?
4. Within multi-site research in pandemics, what responsibilities are there to promote harmonized ethics review, effective monitoring, communication and networking between participating sites?

Case 6.3: The Importance of Effective Research Ethics Review

This case study was written by members of the case study author group.

Keywords Ethics review; Research ethics committee remits and responsibilities; Safety and participant protection; Research quality; Social and scientific value; Multi-centre research; Treatment repurposing

The COVID-19 pandemic brought countless challenges to research ethics committees, including how to deal with a challenging scenario that impacted risk–benefit analyses, and the need to conduct high-quality, rapid analyses of research submitted for ethics review.

A REC received a protocol for a randomized multi-centre clinical trial of a COVID-19 therapy for review. The protocol involved a trial of a therapeutic intervention that was being highlighted in the media and that was also highly politically contentious. In some places, with governmental financial support, the intervention was being provided as part of clinical care for patients diagnosed with COVID-19. The protocol did not contain details about the proposed randomization, nor did it explain what treatment would be used as an adjuvant. There were no references to published studies that would justify the proposal to conduct a Phase III clinical trial. Nevertheless, the project had already been approved by other RECs, despite these methodological flaws.

In addition to the proposal, REC members received information about the institution's strong interest in the management of the research, and about media interest in the intervention. As a result, the REC members considered the possible repercussions of their decision. There were no members of the research team on the REC who could have acted as rapporteurs and provided more information about the study before excluding themselves from the REC's deliberation and decision-making. While analysing the research, the REC identified additional questions relating to the participants' safety, including a lack of procedures for participant monitoring. It also had concerns that the informed consent documentation was inadequate and did not cover relevant areas. Additional concerns arose about whether the study design could answer the research question and whether the principal investigator had an adequate team to carry out the study.

The REC members were divided about whether to approve the research or not. Arguments in favour of approving the research included that the REC should not intervene in the design of the research. Arguments against approving the research included there being insufficient information about the safety of the research and a lack of evidence about the research team's capacity to ensure that risks and burdens of the research could be appropriately managed, and participants adequately protected.

In the context of the COVID-19 pandemic, the REC members felt they had heightened responsibilities, because of the increased vulnerability of the participants, the media attention and the institution's interest and urgency with which it requested a response.

Questions

1. How should a REC manage potential conflicts of interest when there is a very strong institutional interest in undertaking a specific study to address an issue of high priority in a pandemic? Should the institutional interest have been disclosed to the REC? Why?
2. Does the fact that the intervention is already being provided to COVID-19 patients in health-care settings have implications for how a REC should review whether the risks of proposed research are justifiable? Why?
3. Should the REC evaluate methodological flaws in a protocol differently in a pandemic context where there is greater uncertainty and urgency? Why?
4. In a pandemic context where there is intense media interest in potential treatments, should REC evaluations of submitted protocols be influenced by the media? Why?

Case 6.4: Research into the Use of Ozone for Treatment of Patients with COVID-19

This case study was written by members of the case study author group.

Keywords Ethics review; Research quality; Safety and participant protection; Researcher roles and responsibilities; Treatment repurposing; Digital and remote healthcare and research

In the first months of 2020, several clinical teams tried different adjuvant therapies to treat patients with COVID-19. In one study in a country in Latin America and the Caribbean, researchers proposed to provide ozone by direct intravenous administration as a treatment for adult patients with COVID-19. All the procedures, including the preparation and administration of the ozone, would be carried out at the participant's home by a registered nurse. The research protocol was over 150 pages long, with 10 pages assigned to the methodological design, and 23 annexes, including information about the researchers. Two of the three researchers were not registered as clinicians in the country where the study was being conducted.

The research protocol described different types of viruses and the use of ozone as a therapy for Ebola. The protocol noted that ozone had been used in isolated cases of Ebola, rather than being provided as an intervention in a study, and suggested that its use merited further investigation, given the reported recovery after an infection with such a lethal virus. This was the only information provided in the protocol about prior use of ozone as a therapy for a communicable disease in humans. One protocol annex referred to ozone as adjuvant therapy for COVID-19, indicating that the first records of its use should inform further research into its potential as an adjuvant to the treatment (Martínez-Sánchez et al. 2020; Ricevuti et al. 2020).

The inclusion criteria for the study were as follows: patients older than 21 years of age with suspected or confirmed infection with SARS-CoV-2, who did not have symptoms serious enough to require hospital treatment. The proposed research intervention consisted of a calibrated daily intravenous dose of ozone, combined with vitamin C tablets four times a day, for 3 days. According to the protocol, a nurse would carry out all the research processes from a vehicle outside the participant's home. This would include the consent process, taking vital signs, blood samples, possible tests for COVID-19, and the administration of the ozone. The protocol did not adequately detail the biosecurity measures that would be undertaken to avoid infection. Participants would record all their symptoms and respond by telephone to a basic daily questionnaire about their behaviour.

The research ethics committee requested more evidence regarding the safety of the proposed intervention. They pointed to the importance of strict surveillance of ozone administration and additional clinical research into the risks associated with use of ozone therapy when co-morbidities might exist, including chronic diseases. After two rounds of ethics review and requested revisions, the committee continued to have concerns about ethical issues relating to confidentiality, the informed consent process, lack of good clinical practice training, and severe deficiencies in the protocol and its proposed methods. At this point the principal investigator sent a letter questioning the observations and the competence of the committee, as well as stating that the committee review was made "in bad faith".

The committee sent one more letter to the principal investigator to enforce applicable regulations and cite the foundations for its outstanding questions. A few weeks later, the ethics committee accepted a request from the investigator to withdraw the study.

Questions

1. What effect, if any, should the lack of effective treatments in a pandemic context have on the ethics review of studies for a proposed treatment for COVID-19? Why?
2. In a pandemic context where infection-control measures are in place, should such a study be conducted in a person's home and over the telephone? Why?
3. What qualifications should research teams have to conduct clinical research such as this? Should that vary based on where the study is being conducted and whether it addresses questions relevant to an outbreak or epidemic? Why?
4. Should the research ethics committee have rejected this proposal from the outset? Why?

References

Martínez-Sánchez, G., A. Schwartz, and V.D. Donna. 2020. Potential cytoprotective activity of ozone therapy in SARS-CoV-2/COVID-19. *Antioxidants* 9(5): 389. https://doi.org/10.3390/antiox9050389.
Ricevuti, G., M. Francini, and L. Valdenassi. 2020. Oxygen-ozone immunoceutical therapy in COVID-19 outbreak: Facts and figures. *Ozone Therapy* 5: 914–917.

Case 6.5: Reviewing the Use of Convalescent Plasma

This case study was written by members of the case study author group.

Keywords Ethics review; Research ethics committee remits and responsibilities; Research quality; Social and scientific value

At the beginning of the COVID-19 pandemic, to enhance research oversight response capabilities, the health authorities of a country in Latin America created an ad hoc national research ethics committee (NREC) responsible for reviewing all COVID-19 research to be conducted in the country.

During early stages of the pandemic, many studies around the world started testing convalescent blood plasma as a potential treatment for COVID-19 because there was evidence it had improved outcomes in patients with other diseases, such as Influenza A and Ebola, and because of its relatively low risk profile. In May 2020, the first protocol in the country aimed at determining the efficacy of convalescent plasma for COVID-19 was submitted to the NREC for review. In the following months five more protocols proposing to study the safety and efficacy of convalescent plasma for COVID-19 were submitted. All of them studied the same population (hospitalized moderate to severely ill patients); however, the study endpoints were diverse and the characteristics of the plasma used (e.g. its antibody level) were not standardized, which hindered comparison of findings across trials.

The first protocol, which was approved by the NREC, was a single-site, open-label, non-randomized trial with a sample of fewer than 100 participants. The second and the third studies, which were also approved by the committee, were multi-site randomized controlled clinical trials with samples of more than 200 participants. The remaining three studies were open-label and non-randomized clinical trials, and together sought to enrol approximately 200 participants. The members of the NREC deliberated about the seemingly lower scientific strength of these last three trials, the fact that they enrolled similar populations, and the implications of this for their review. Some NREC members argued that they should go on approving trials testing convalescent plasma in similar populations even if they were of lower scientific quality than the second and third clinical trials, which had already been approved. Ultimately, they decided that the urgency of advancing research to generate knowledge to address COVID-19 justified approving all the studies, even if they would produce weak evidence.

A few months later, the media reported that in high-income countries convalescent plasma was being provided to patients through expanded access or compassionate use programmes, on the basis of the modest evidence suggesting a positive benefit to risk ratio, given the lack of effective treatments for COVID-19 at the time. Taking note of this, people in the country started demanding that convalescent plasma be made available to COVID-19 patients outside the research context. Since the country did not have ethics or regulatory guidance for emergency use of unproven and unregistered interventions, convalescent plasma was given to some COVID-19 patients outside of a study protocol. In a few instances, clinicians requested that the NREC review these proposed therapeutic uses, and some members argued that the committee should provide oversight, because the use of unproven interventions outside of research should adhere to ethical standards. However, the NREC ultimately decided not to review emergency use applications, on the grounds that this fell outside its mandate.

Questions

1. In the pandemic context was it ethically justifiable for the NREC to approve all of the clinical trials testing convalescent plasma, considering their different scientific strengths? If so, why? If not, on what grounds should they have not approved certain studies?
2. Would it have been acceptable for the NREC to withdraw their approval of the first protocol, following approval of other protocols with stronger social and scientific value? Why?
3. Taking into account that all the trials were testing the same intervention in similar populations, what role, if any, should the NREC play fostering collaboration and coordination among the research teams to accelerate or enhance the research? Should the committee establish procedures to anticipate upcoming submissions, or is that beyond its mandate? Why?
4. Given the lack of ethics and regulatory guidance in the country regarding the emergency use of unproven interventions outside of research, should the NREC have reviewed the submitted request for therapeutic use outside a protocol? If so, are there any ethical issues that the NREC should consider in light of the potential for simultaneous use of unproven interventions outside of research and enrolment into clinical trials of those interventions?

Case 6.6: A Phase III COVID-19 Vaccine Trial

This case study was written by members of the case study author group.

Keywords Ethics review; Research ethics committee remits and responsibilities; Regulatory review; Vaccines; Emergency Use Authorisation; Multi-centre research; Placebo control

In January 2021, there was no effective anti-viral treatment for COVID-19. A mainstay of addressing both mortality and morbidity due to the disease is to vaccinate health-care workers, frontline workers and high-risk groups, including the elderly and those with co-morbidities. Initial vaccine development primarily occurred in high-income settings with relevant expertise and technology. By late 2020 many vaccines had begun receiving emergency use authorization (EUA) and several countries had begun rolling out vaccines for their populations. Some of the initial vaccines deployed after EUA included Vaxzevria, an adeno-virus vector vaccine developed by the University of Oxford and AstraZeneca, and the Pfizer Bio NTech and Moderna vaccines (both mRNA vaccines).

These vaccines were also sought by governments in low- and middle-income countries (LMICs) to use for their population, regardless of whether the vaccine was originally developed or tested in their country. Despite their approval elsewhere, in such circumstances, national regulatory agencies in LMICs often asked for a small bridging trial to ascertain safety and generate limited efficacy data to enable the regulator to grant an EUA. These studies were usually multi-centre, so as to be completed in the shortest possible time.

One vaccine manufacturer in a LMIC partnered with a vaccine manufacturer from a HIC to produce a vaccine which had received an EUA, using the same technology as the HIC manufacturer. A multi-centre placebo-controlled trial was planned with the approval of the drug regulator in October 2020 to ascertain safety and generate limited efficacy data before distributing the vaccine more widely. A key exclusion criterion in the trial was the presence of antibodies to SARS Co-V-2 indicating a past infection.

At their discussion of the ethical acceptability of the proposal, one of the members of a local research ethics committee argued that approval should be conditional on the national pharmaceutical company:

(a) providing the vaccine free of cost to all those who were in the placebo arm
(b) providing the vaccine free of cost to all the employees of the institution at the site where the study would be carried out, and
(c) providing the vaccine at very low cost on a not-for-profit basis to the nation's population once the EUA was received.

Questions

1. Is it ethically acceptable to conduct bridging trials of this type for vaccines that have already received an EUA elsewhere, if doing so will delay the introduction of the vaccine in a population? Why?
2. How do you think the researcher should respond to the ethics committee comments about the responsibility of the pharmaceutical company? Why?
3. If the trial did go ahead, what key topics which should have been addressed in the informed consent process?

Case 6.7: Research on Teleconsultation

This case study was written by members of the case study author group.

Keywords Regulatory review; Safety and participant protection; Research priority setting; Researcher roles and responsibilities; Digital and remote healthcare and research; Qualitative research; Researcher safety

During the COVID-19 pandemic, teleconsultations have become the preferred mode of delivery of health care, particularly when monitoring patients with chronic diseases such as diabetes, who require frequent check-ups and are at high risk of developing severe complications if they contract COVID-19. Patients can have teleconsultations with their health-care providers without having to visit the clinics or hospitals, thereby reducing their risk of exposure to infection. However, the effectiveness and safety of teleconsultations for monitoring patients with diabetes, who often also have other comorbidities, remain uncertain, especially in a country where teleconsultation is still relatively new.

In an Asian country, Professor A. and her team, who are responsible for diabetes care in their hospital, were keen to explore the challenges faced by doctors and patients when teleconsultations were used to monitor diabetes patients in a hospital diabetes clinic. They planned to interview both doctors and patients about the challenges posed by teleconsultations, as well as perusing the teleconsultation records to examine and understand the interactions. The team decided to seek the consent of the doctors and patients to take part in research interviews and share electronic medical records and teleconsultation records. The researchers planned to seek consent both verbally via telephone and electronically via e-consent, hoping that this would avoid the need for the participants to give their consent in person. The research project was approved by the hospital research ethics committee, which agreed with the research team that remote consent was appropriate in order to protect both patients and researchers from exposure to the SARS-CoV-2 virus at the hospital.

However, the research team faced an obstacle when applying to access the hospital's electronic medical records, which were used for documenting patient teleconsultations. The country's Telehealth Act, which was enacted 20 years ago but never formally enforced, does not permit clinical consultations to take place without a face-to-face consent process. The hospital's legal adviser insisted that research participants needed to give written consent for researchers to access their electronic records during a face-to-face meeting. Despite being given an explanation of the potential benefits of teleconsultations during the pandemic, as well as the potential risks to patients and researchers of face-to-face consent, the hospital's legal adviser insisted that the hospital would not take the risk of going against the law.

Consequently, Professor A. could obtain face-to-face written consent only from patients who came to the hospital for routine blood tests before their visits to the diabetes clinic. The progress of the research was slowed down significantly by slow recruitment as fewer patients attended the clinic since many of them were fearful about entering the hospital. Also, new Standard Operating Procedures during the pandemic meant that fewer appointments were scheduled each day. Professor A. and her team saw a reduction of about 70% in patients attending the clinic for regular blood tests. Professor A. also felt that it was unethical to expose the research assistants to the risk of catching COVID-19 when they sought to obtain consent from the patient in person.

Questions

1. What might be appropriate responses to institutional policies or legal regulations which appear to conflict with the changes in research practice proposed to enable research to be conducted ethically during a pandemic?
2. How should the hospital's research ethics committee respond in a pandemic to the requirement to seek face-to-face consent, given the risks of infection to both patients and researchers?
3. What are the ethical implications of not conducting research into the safety and efficacy of rapid changes in clinical practice introduced in response to the pandemic (such as the use of teleconferences to aid patient monitoring)?

References

Carracedo, S., A. Palmero, M. Neil, A. Hasan-Granier, C. Saenz, and L. Reveiz. 2020. The landscape of COVID-19 clinical trials in Latin America and the Caribbean: Assessment and challenges. *Revista Panamericana de Salud Publica* 44: e177. https://doi.org/10.26633/RPSP.2020.177.

CIOMS. 2016. *International ethical guidelines for health-related research involving humans.* Geneva: Council for International Organizations of Medical Sciences. https://cioms.ch/wp-content/uploads/2017/01/WEB-CIOMS-EthicalGuidelines.pdf.

Herper, M., and E. Riglin. 2020. Data show panic and dis-organization dominate the study of Covid-19 drugs. *STAT News*, July 6. https://www.statnews.com/2020/07/06/data-show-panic-and-disorganization-dominate-the-study-of-covid-19-drugs/.

ICMR. 2020. *National guidelines for ethics committees reviewing biomedical research during COVID-19 pandemic.* New Delhi: Indian Council of Medical Research. https://ethics.ncdirindia.org//asset/pdf/EC_Guidance_COVID19.pdf.

London, A.J. 2021. Self-defeating codes of medical ethics and how to fix them: Failures in COVID-19 response and beyond. *American Journal of Bioethics* 21(1): 4–13. https://doi.org/10.1080/15265161.2020.1845854.

London, A.J., and J. Kimmelman. 2020. Against pandemic research exceptionalism. *Science* 368(6490): 476. http://science.sciencemag.org/content/368/6490/476.

PAHO. 2016. *Zika ethics consultation: Ethics guidance on key issues raised by the outbreak.* Washington, DC: Pan American Health Organization. https://iris.paho.org/handle/10665.2/28425.

———. 2020a. *Emergency use of unproven interventions outside of research. Ethics guidance for the COVID-19 pandemic.* Washington, DC: Pan American Health Organization. https://iris.paho.org/handle/10665.2/52429.

———. 2020b. *Ethics guidance on issues raised by the novel coronavirus disease (COVID-19) pandemic.* Washington, DC: Pan American Health Organization. https://iris.paho.org/handle/10665.2/52091.

———. 2020c. *Guidance and strategies to streamline ethics review and oversight of COVID-19-related research.* Washington, DC: Pan American Health Organization. https://iris.paho.org/handle/10665.2/52089.

———. 2020d. *Guidance for ethics oversight of COVID-19 research in response to emerging evidence.* Washington, DC: Pan American Health Organization. https://iris.paho.org/handle/10665.2/53021.

———. 2020e. *PAHO does not recommend taking products that contain chlorine dioxide, sodium chlorite, sodium hypochlorite, or derivatives.* Washington, DC: Pan American Health Organization. https://iris.paho.org/handle/10665.2/52515.

———. 2020f. *Template and operational guidance for the ethics review and oversight of COVID-19-related research.* Washington, DC: Pan American Health Organization. https://iris.paho.org/handle/10665.2/52086.

———. 2022. *Catalyzing ethical research in emergencies. Ethics guidance, lessons learned from the COVID-19 pandemic, and pending agenda.* Washington, DC: Pan American Health Organization. https://iris.paho.org/handle/10665.2/56139.

Palmero, A., S. Carracedo, N. Cabrera, et al. 2021. Governance frameworks for COVID-19 research ethics review and oversight in Latin America: An exploratory study. *BMC Medical Ethics* 22. https://doi.org/10.1186/s12910-021-00715-2.

WHO. 2016. *Guidance for managing ethical issues in infectious disease outbreaks.* Geneva: World Health Organization. http://apps.who.int/iris/bitstream/10665/250580/1/9789241549837-eng.pdf.

————. 2020a. *Ethical standards for research during public health emergencies: Distilling existing guidance to support COVID-19 R&D*. Geneva: World Health Organization. https://apps.who.int/iris/handle/10665/331507.

————. 2020b. *Guidance for research ethics committees for rapid review of research during public health emergencies*. Geneva: World Health Organization. https://www.who.int/publications/i/item/9789240006218.

Chapter 7
Ethical Issues Associated with Managing and Sharing Individual-Level Health Data

Sharon Kaur and Phaik Yeong Cheah

Abstract The COVID-19 pandemic has resulted in the generation of an unprecedented and exponentially mounting volume of data, including individual-level health data, bringing into sharp focus the importance of thinking about what constitutes ethical use of data in a public health emergency. The timely and appropriate use of such data (e.g. data from public health surveillance, electronic health records and research projects) has great potential to contribute to successful public health policies, effective therapeutic interventions and enhanced public support for, and trust in, governmental responses to the pandemic. However, a number of ethical issues arise from the use of different kinds of data, and the ways in which they are collected, processed and shared in the context of research during a pandemic. Two broad principles are generally associated with managing and sharing health data in research: first, that researchers should ensure research is carried out in a way that is respectful of persons and communities; and second, that the research is carried out in a manner that is fair to stakeholders, i.e. that it promotes equity. These should also remain the foundational principles of data sharing during a public health emergency. The principle of respect for persons and communities requires careful attention to be paid to consent processes for data sharing, justifications for waiving consent and approaches to protecting privacy and confidentiality. The promotion of equity prompts consideration of how the needs of differing stakeholders in data sharing are recognised and balanced, including appropriate forms of recognition for data sharers, and fair benefit sharing with the individuals and communities data have been collected from. The cases in this chapter illustrate issues arising when populations contribute data to a symptom-checker app, when heightened concerns arise raised

S. Kaur (✉)
Faculty of Law, Universiti Malaya, Kuala Lumpur, Malaysia
e-mail: kaursh@um.edu.my

P. Y. Cheah
Mahidol Oxford Tropical Medicine Research Unit, University of Oxford, Oxford, UK

Mahidol Oxford Tropical Medicine Research Unit, Mahidol University and Nuffield Department of Medicine, University of Oxford, Oxford, UK

© PAHO and Editors 2024
S. Bull et al. (eds.), *Research Ethics in Epidemics and Pandemics: A Casebook*,
Public Health Ethics Analysis 8, https://doi.org/10.1007/978-3-031-41804-4_7

about privacy and confidentiality in the context of collecting data about individuals who are potentially easily identifiable by their demographic characteristics, when very sensitive data is collected, and when a waiver of consent to access survey data is requested to enable potential participants of a study to be identified and contacted.

Keywords COVID-19 pandemic · Research ethics · Public health emergencies · Data protection, access and sharing · Privacy and confidentiality · Boundaries between research, surveillance and clinical care · Digital and remote healthcare and research · Consent · Researcher roles and responsibilities · Ethical review · Vulnerability and inclusion · Consent · Ethical review

7.1 Introduction

The COVID-19 pandemic has brought into sharp focus the importance of thinking about what constitutes ethical use of data in a public health emergency. This is because the pandemic has resulted in the generation of an unprecedented and exponentially mounting volume of data, including individual-level health data. The timely and appropriate use of such data (e.g. data from public health surveillance, electronic health records and research projects) has great potential to contribute to successful public health policies, effective therapeutic interventions and enhanced public support for, and trust in, governmental responses to the pandemic (Han et al. 2021; Rios et al. 2020). However, the collection and use of individual-level health data may also be problematic if not managed appropriately. Concerns have been raised about privacy and data protection (Ienca and Vayena 2020). These are particularly relevant in certain situations: first, in relation to the use of surveillance and registry data and the extent to which reporting individual-level health data may violate trust and create fear of stigmatization and discrimination (Bayer and Fairchild 2000); and second, in the context of the collection and curation of data (particularly when there are concerns about data being sensitive) by novel and remote platforms (Newlands et al. 2020). In both these instances, the apparent loss of individual control over the collection and use of personal health data is seen to strike at the heart of values such as autonomy, privacy and trust (Vayena and Blasimme 2017).

In the context of conducting research during the pandemic, issues relating to the use and sharing of individual-level health data have taken on increased urgency. There have been calls for greater use of transparent and open-science data-sharing options, and speedier sharing of data to inform COVID-19 patient management and response (Moorthy et al. 2020; Homolak et al. 2020). International research funders such as the Bill and Melinda Gates Foundation and Wellcome have mandated that their grantees share data from research related to COVID-19 as soon as the data collection is completed, regardless of publication status. International expert and working groups have been established to facilitate effective, equitable and ethical data-sharing (Fegan et al. 2021).

Most countries have laws that regulate the processing of personal information, which include the managing and sharing individual level health data. The European General Data Protection Regulation (GDPR) (2016) is an example of such a document. Researchers should be mindful of the legal duties imposed by such regulations. However, almost all of these legislative frameworks are likely to provide exemptions for the processing of data for special reasons or if they fall within special categories such as reasons of public interest relating to public health (for example, Article 9(2)(i) of the GDPR).

As demonstrated by the cases in this chapter, a number of ethical issues arise from the use of different kinds of data, and the ways in which they are collected, processed and shared in the context of research during a pandemic. Two broad principles are generally associated with managing and sharing health data in research: first, that researchers should ensure research is carried out in a way that is respectful of persons and communities; and second, that the research is carried out in a manner that is fair to stakeholders, i.e. that it promotes equity. Although the issues raised in these cases engage the same broad ethical principles as research in non-emergency settings, two observations can be made. First, the principles play out against a very different background during a pandemic, when the value of data may be perceived quite differently. The literature suggests that the value of collecting, processing and sharing data during a pandemic is most often linked to the utility of the data, which is commonly measured by assessing potential benefits in terms of (1) the reduction of suffering of current and future populations, (2) improvement of quality of life during and after the pandemic and (3) the reduction of the socio-economic impact of the pandemic (Bull et al. 2015b; Pratt and Bull 2021). Whether and in what way the approaches to maximizing the utility of data during a pandemic might be justified, or come into conflict with the traditional ethical principles noted above, will be considered by reference to the case studies in this chapter. Second, these cases highlight the importance of context and the need to recognize the salient considerations that should inform how the two broad ethical principles of specific relevance – respect for persons and communities, and promoting equity – should be taken into account during a pandemic.

7.2   Respect for Persons and Communities

It is important that researchers and other secondary users of data respect the persons and communities they engage with. The principle of respecting persons recognizes that individuals should be treated as autonomous agents, and that persons with diminished autonomy are entitled to protection (National Commission 1979). As autonomous agents, individuals and communities have the right to make their own decisions about how their information is collected and used. Respecting persons and communities means that individuals must provide consent to the use of their data, and that their privacy is assured. There are certain situations where waivers of consent are permitted provided that certain conditions are met (CIOMS 2016). But even in such situations, researchers are required to consider whether the consent process might be modified in order to preserve as much of the individual's autonomy as possible.

7.2.1 The Consent Process

A properly executed consent process is a vital aspect of respecting individual research participants. As part of such a process, adequate information is provided to participants, and participants understand what is proposed, including the nature of any risks and benefits to them and how such risks are to be managed and minimized. Moreover, this consent should be voluntarily given and researchers should ensure that there is no undue influence or deception involved. Participants should also be aware that they may withdraw from the research at any time without the need to provide any explanation or justification, and without any penalty or prejudice to any treatment they may be receiving (CIOMS 2016).

Researchers engaged in collecting, processing and sharing health data should ensure that effectively designed consent processes are put into place. Participants should be provided with information that is explained in a way that allows them to have an appropriate level of understanding. This information should be sufficient and relevant. Participants should understand what health data will be collected and for what purposes, and how their data will be stored and shared. Researchers must carefully consider the risks to participants, and, where relevant, to their communities. In cases where potential participants may have low literacy levels, be unfamiliar with health research, and suffer from social vulnerability, extra measures may need to be taken to ensure that they are able to provide meaningful consent (Bhutta 2004; Cheah et al. 2018).

Given the complexity and abstract nature of data collection, use and processing, particularly in the context of mobile applications and novel platforms, and given most people's unfamiliarity with it, providing accessible information about data-sharing can be challenging. Case 7.1 illustrates this with an example of a mobile application that collects both sensitive personal data and other personal data. The data collected and processed by the application are used and shared with different parties in accordance with the legal restrictions in place and on the privacy policy of the developers. The envisaged creation of a consortium and data pool for research purposes will mean that the application will be used both as a source of data for research and as a tool for individuals to assess their symptoms. The many different activities associated with the application, as well as the many different ways in which data may be processed and shared, prompt consideration of researchers' obligations to design a consent process that ensures participants understand how their data may be used and shared in the context of research and what measures will be taken to protect their privacy.

Significantly, during a public health emergency, research takes on a new urgency, and the extensive collection, processing and sharing of data can be particularly valuable. In relation to the ethical principle of requiring meaningful consent from participants, this raises the challenge of whether and how the prioritization of research during a pandemic should be balanced against traditional consent mechanisms. In such situations, modifications of the consent process may be permitted if there is no other feasible or practicable option, the research has important social value and it poses no more than minimal risks to participants. This will be explored in Cases 7.1 and 7.2 in relation to two aspects of consent: consent for future use of data and approaches that involve waiving consent or opting out of sharing.

7.2.2   Consent for Use of Data in the Future

Case 7.1 envisions a variety of future uses and sharing of data, which then raises the issue of what sort of consent would be appropriate in such a case. There has been much debate about appropriate models of consent to allow sharing, storage and future use of data. There is a spectrum of approaches to consent for the future use of data. It ranges from "specific consent" (where the participant would be re-contacted for permission in connection with any future research study) to "blanket consent" (where any use is permissible, including uses unrelated to health) to no consent at all (Tindana et al. 2019). Between the two extremes, there has been increasing support for "broad consent", which is consent for unspecified future use as long as the future use is within the scope of the broad consent, for example "health research" or "malaria research", and with appropriate governance processes in place (CIOMS 2016). However, some authors have challenged the concept of "broad consent", asking whether it can constitute informed consent (Sheehan 2011). If the project in Case 7.1 has not set a time limit on how long it will hold participants' information and what sort of research might be carried out (given the value of the data in the time of COVID-19) the researchers may be requesting blanket consent.

In considering whether "blanket consent" for future use of health data in relation to Case 7.1 might be justified during a pandemic, it may be worth taking note of certain factors. The first is the utility of digital symptom trackers as a public health tool. The literature suggests that symptom trackers are very valuable tools for monitoring a public health threat and enabling a quick response. They also provide governments with information to assist them in allocating resources and generally minimizing or controlling outbreaks (Gasser et al. 2020). From a public health perspective, the use of blanket consent might be the most efficient way of maximizing the utility of such a tool, and careful consideration is required of how the interests of participants should be balanced with threats to public health and safety in a pandemic.

7.2.3   Waiving Consent and the Opt-Out Approach

Case 7.2 considers whether waivers of consent and opt-out approaches are appropriate in the context of a pregnancy outcomes registry. Waivers of consent must typically be approved by research ethics committees (CIOMS 2016) and are approved only when specific conditions are met, such as "the waiver or alteration will not affect the rights of the subjects, the research cannot be carried out without the waiver, and, when appropriate, subjects will be provided information after participation" (Berg et al. 2018). However, the commentary to Guideline 10 of the Council of International Organizations of Medical Sciences (CIOMS) guidelines provides special considerations for waiving informed consent in studies using data from health registries, citing the importance of having comprehensive and accurate information about an entire population, the avoidance of undetectable selection bias and the need to equitably distribute benefits and burdens across a population (CIOMS 2016).

Significantly, in relation to public health emergencies, the guidance seeks to balance the social value of such research against the risks of violating individual autonomy. First it recognises in the commentary that when such studies are conducted under public health mandates or by public health authorities, no ethical review of waiver of consent is needed as the research is mandated by law. However, consent cannot be waived by a public health authority if the research combines data in a registry with new activities that involve direct contact with participants. Moreover, even in such situations, it stipulates that when the use of such data no longer constitutes a public health activity, researchers must seek individual consent or obtain ethics review approval to waive consent under the conditions in Guideline 10 (CIOMS 2016). It is important to recognise that the ethical justification for a public health mandate is a separate question entirely. Whether or not any particular public health mandate is ethical will depend on a separate ethical analysis based on principles related to public health ethics such as the harm principle, least coercive or restrictive means, reciprocity principle and transparency principles (Upshur 2002). Second, in the absence of a legal mandate, when considering whether to waive individual consent, researchers and research ethics committees are required to consider whether there are any other modifications that can be made to the informed consent process that would allow for the greatest expression of individual autonomy (CIOMS 2016).

In this case, researchers chose an opt-out approach, which on the face of it appears appropriate and respectful of the rights of patients. According to CIOMS (2016), an opt-out procedure must fulfil the following conditions: (1) participants need to be aware of its existence; (2) sufficient information needs to be provided; (3) participants need to be told that they can withdraw their data; and (4) a genuine opportunity to object has to be offered. These requirements may be challenging to fulfil in research conducted with disadvantaged or marginalized groups (CIOMS 2016). Despite the fact that this registry was established in a high-income country, information was collected from staff providing maternity care across a range of health settings and would have very likely included data about women from disadvantaged and marginalized groups. Significantly, in Case 7.2, despite the fact that none of the women opted out of the research, fewer than 10 cases were added in the early months of the pandemic. Whether this was attributable to an effective national COVID-19 response or the reluctance of health providers to register women, it limits the social value of the research and the utility of the dataset. In the context of a public health emergency, researchers and ethics committees may need to give particular regard to whether maximizing the utility of a pregnancy outcomes registry may warrant a waiver of consent.

7.2.4 Privacy

Protecting the privacy and confidentiality of participants remains one of the main concerns of the storage, access to, management of and sharing of data in research (Bull et al. 2015a). Custodians of data are required to make arrangements to protect the confidentiality of the information linked to the data and to limit access to the material relating to third parties (CIOMS 2016). If there is a breach in confidentiality, participants may risk being blamed or stigmatized, and public trust in science and research may be undermined. In the context of COVID-19, for example, some COVID-19 patients have had to respond publicly to allegations about non-compliance with public health measures and defend themselves (Atan 2020). In such contexts people with symptoms of COVID-19 may not want to come forward and get tested. There are two important privacy-related issues that arise from the cases in this chapter: the issue of de-identification of data and the use of existing datasets.

7.2.5 Curation, De-identification and Anonymization

Careful curation and de-identification of data is frequently offered as a way of protecting individual anonymity, but some have argued that identifiability exists on a continuum (Rothstein 2010). Although personally identifying information, such as name, address and date of birth, are omitted when data are shared, the data may still have identifying characteristics. Low numbers of enrolled patients may also heighten the risks of identification, as noted in Case 7.2. In addition, data scientists have proved on multiple occasions that datasets that were thought to be anonymized could be linked with other public health data to identify the specific data subject (Sweeney 2000). With the advent of Big Data and an increasing move to link large databases and permit exploration with machine learning and artificial intelligence approaches, it may become increasingly difficult to ensure the anonymity of individuals, and researchers should be aware of the risks to both individuals and communities in the event of re-identification. This is a particular concern in relation to the sharing of health records, and the lack of consistency in the applications of anonymization remains an unresolved issue.

It is also important to bear in mind the possibility of harm to groups or communities in the form of stigma or discrimination. Although data are de-identified at the individual level before they are shared, the dataset might still be attributable to a certain community, and risks of stigmatization or discrimination should be taken into account. This could impact employment opportunities or lead to discrimination by insurance companies (Rothstein 2010).

Sharing qualitative data poses an additional set of challenges. This is especially true when the data collected are derived from individuals who are easily identifiable by their profession, location, idiolect or opinions (such as in Case 7.3) and/or address intimate aspects of people's lives and are considered very sensitive (as in Case 7.4). It is generally agreed that the more potentially identifying information is removed from a dataset, the less useful such data may be. The ethics committees in Cases 7.3 and 7.4 flag this very issue, and are concerned about preserving the confidentiality of participants.

Cases 7.2–7.4 raise concerns about possible re-identification of participants, and it is worth exploring who these participants are in each case and whether in the context of a pandemic, it may be ethical to proceed with research despite the possible risk of identification. Researchers and ethics committees will need to identify and balance the potential risks to the specific group of participants against the anticipated utility of the data in relation to answering important questions related to the public health emergency.

7.2.6   Use of Existing Datasets

Challenges arise when researchers seek to use archived data from prior research, clinical care or other public health activities without having obtained informed consent from participants for their future use. An example of this type of study is a review of old hospital records, where participants have not been asked if they consent to their data being used for research purposes in the future. Another example is the use of previously collected datasets generated for another purpose, as in Case 7.4, where the proposal was to use health data from earlier surveys to identify potential participants for proposed research. In such cases, the CIOMS 2016 guidelines state that "the research ethics committee may waive the requirement of individual informed consent if: (1) the research would not be feasible or practicable to carry out without the waiver; (2) the research has important social value; and (3) the research poses no more than minimal risks to participants or to the group to which the participant belongs" (CIOMS 2016). In Case 7.5, researchers were proposing to contact some participants in the survey to invite them to participate in a new study, but participants had not provided consent to be contacted. The proposed study could directly address their health needs. In the context of Case 7.5, it seems that criteria 1 and 2 may be met. As for criterion 3, researchers should find ways to minimize any risks to participants, including when contacting them to invite them to take part in a study. It is important to bear in mind that some of these risks, such as being stigmatized, may not be obvious to the researcher.

7.3 Promoting Equity

It is important that primary and secondary users of data ensure that approaches to data processing and use are equitable (GLOPID-R 2018). They should recognize and balance the needs of the different communities involved. This includes those who collect and generate the data, secondary users of the data, the individuals and communities from which the data originate, and the funders of the collection effort (Vayena and Blasimme 2017). Sharing data widely or allowing completely open access to data, with minimal governance mechanisms and oversight, has previously generated significant concerns related to equity, such as lack of recognition of the efforts of data generators, inequitable access to data, and failure to ensure fair benefits to study participants and communities, especially when dealing with data collected from potentially vulnerable populations (Pratt and Bull 2021).

Researchers working in low-resource settings have raised the issue that data-sharing has the potential to exacerbate existing inequalities between health researchers working in low-resource and high-resource settings (Serwadda et al. 2018). They worry that they are reduced to data collectors in the service of highly skilled data analysts and statisticians from high-resource settings. Some approaches that can be taken to prevent this from happening include sharing credit for scientific outputs, ensuring that capacity-building measures are built into research proposals, and ensuring that collaborations are negotiated on the basis of mutually beneficial data-sharing arrangements. Unfortunately, time and resources are often scarce during a pandemic, when the research imperative takes on a new urgency and some of these approaches may not be feasible. For instance, the national research ethics committee in Case 7.3 is described as unable to function optimally, owing, among other things, to lack of resources and inadequate training. A pandemic may not be an ideal time to focus on building the capacity of existing ethics committees when research needs to be carried out in a timely fashion. In such situations, it is suggested that other organizations may need to provide assistance to local research ethics committees to help them overcome these challenges (Smith and Upshur 2019).

Participants and communities involved in research have a valid interest in experiencing the benefits of research arising from the use of their data. In the context of low- and middle-income countries and vulnerable populations, particular attention should be paid to research projects that rely on technologies that are not accessible by or available to certain populations. Fair access to the benefits of research should require that individuals or communities are not excluded from the potential benefit of participating in research because of a digital divide. These concerns may be relevant to Case 7.1, which involves the use of a mobile application developed and deployed in Europe and the Americas. Even in high-income countries, researchers should be aware of the need to ensure that vulnerable and marginalized communities are not excluded from research that could confer benefits to them. Pandemics have historically affected disadvantaged communities very differently, with higher rates of infection, mortality and morbidity (Bambra et al. 2020; Osterrieder et al. 2021; Schneiders et al. 2022), and depriving these communities the benefits of research is unethical.

However, it is not always clear what would constitute a benefit and who it should be shared with. Stakeholders have discussed the importance of both direct and indirect benefits (Bull et al. 2015a). Indirect benefits are particularly relevant in the context of secondary research, which may not address health issues of relevance to participants and communities.

It is also worth noting that the sharing of data for commercial purposes may be a sensitive issue, particularly in relation to a public health emergency (Pratt and Bull 2021; Tangcharoensathien et al. 2010; Ghafur et al. 2020). Community expectations and views may also vary considerably, depending on historical, political and cultural contexts, and researchers should be mindful of the interests of communities when sharing data in such contexts. The involvement of commercial interests in research is also seen as potentially problematic, as they may inhibit timely sharing of data and results as well as being reluctant to share negative data and results. Data-sharing procedures should be agreed in advance to ensure timely access to data and results (including negative ones) (GLOPID-R 2018).

7.4   Conclusion

The collection, processing and sharing of individual-level health data are a critical part of public health emergency responses. Timely and effective access to and analysis of data in health research can generate a "deeper understanding of an outbreak, its impact on patients, and effective methods of control – supporting more effective public health responses" (GLOPID-R 2018). However, the collection, processing and sharing of data must be done in an ethically appropriate manner. The broad ethical principles of respecting persons and communities, and promoting equity, which apply to the use of health data in non-emergency research should also remain the foundational principles of data sharing during a public health emergency. However, during a pandemic there are salient considerations that should inform how these broad principles should be applied on a case-by-case basis, as demonstrated above.

Case 7.1: A Multinational COVID-19 Symptom Checker Application

This case study was written by members of the case study author group.

Keywords Boundaries between research, surveillance and clinical care; Privacy and confidentiality; Data protection, access and sharing; Consent; Digital and remote healthcare and research; Citizen science

A growing number of technological initiatives have emerged to surveil, predict and control the spread of COVID-19. These include a class of tools known as symptom checkers, which have been an important part of the COVID-19 pandemic response. At their most basic level, symptom checkers prompt users to enter their symptoms (e.g. fever, cough, loss of smell, shortness of breath) to help identify possible causes and treatments. More sophisticated versions may also provide triage decisions, including recommending that patients seek medical attention and/or diagnostic testing (see Miller 2015; Berry 2018).

The COVID-19 symptom checker in this case was developed by a commercial health science company in collaboration with researchers at several hospitals and academic institutions. In addition to the above functions, this particular symptom checker asks users to provide health-related information (e.g. age, height, weight and sex at birth) and record relevant health information (e.g. symptoms, COVID-19 test results, treatments and pre-existing conditions) daily. The data are made available to the app developers and their partners, who intend to use it to advance scientific research, for instance through refining symptom recognition and identifying high-risk geographic areas and characteristics of individuals.

Available as a mobile application in app stores in Europe and the Americas, the symptom checker has been downloaded by several million individuals. It collects both sensitive personal data and other personal data. This self-reported information is combined with software algorithms to predict who has had the virus and to track COVID-19 infections. As outlined in the symptom checker's privacy policy, the data are protected by the European Union's General Data Protection Regulation (GDPR) and can only be utilized for medical science and to assist the government body responsible for the nation's health-care system. As this is a not-for-profit initiative, data cannot be used for commercial purposes (i.e. sold).

According to this privacy policy, anonymized data may, however, be shared with research institutions beyond the project's partners. These encompass hospitals, clinics, universities, health charities and government actors. Certain personal data are also shared with third-party processors, including analytics, hosting, communications, security and fraud prevention services (e.g. Google Analytics, Amazon Web Services or MailChimp). These parties process some of the users' personal data on behalf of the company, but are unable to utilize the data for their own purposes. As some of these institutions and processors reside in North America, they are not as readily governed by the GDPR. The project has not set a time limit on how long it will hold participants' information, which, the developers state, is due to the value of such data for researchers studying both COVID-19 and epidemic spread more generally.

The symptom checker team reached out to investigators of cohort and clinical studies to offer this tool (including the potential for customization) at no cost. As part of this outreach, the team publicized its intent to create a consortium and pool data for research purposes. If app users are already part of an existing research study or trial, they can request that their data from the app be shared with investigators on that study. Data collected via the app – including through lifestyle surveys – have been utilized in several preprints and published papers.

As stated in the project's terms of use, the app does not offer medical advice and is not meant to diagnose or treat any conditions. The project's website and communications emphasize the app's potential to support collaborative COVID-19 research and contribute to the fight against this novel disease. As both a source of data for researchers and a tool through which individuals can monitor their symptoms, the project has been endorsed by a range of state actors, health-based charities and doctors' membership bodies. Users can add multiple profiles and report symptoms on behalf of others. For example, parents can log their children's health. The application offers a "daily insights" programme for registered schools, where students' information is anonymized and aggregated. This programme aims not only to support decision-making by tracking how many children are unwell in a particular school network, but also to further general understanding of COVID-19 in children.

The project also runs a vaccine and trial registry and has communicated to app users its intentions to prioritize regular users of the symptom checker when providing information about vaccine trials and other preventive treatments. Individuals can join the project's mailing list to learn more about future studies. The application is free and the project plans to use donations and grants to cover its costs.

Questions
1. Which of the above activities should be classified as research (or not research)? Why?
2. Should this app secure consent to data-sharing from users? Why? If so, what consent should be sought and what would be needed to achieve this?
3. What ethical considerations should guide (a) how data collected by the app should be used and (b) with whom it should be shared?
4. Should there be a time limit on how long these data can be held and/or used? What factors should determine this decision?

References

Berry, A.C. 2018. Online symptom checker applications: Syndromic surveillance for international health. *Ochsner Journal* 18(4): 298–299.

Miller, J. 2015. Checking up on symptom checkers. *Harvard Medical School News and Research.* Blog. https://hms.harvard.edu/news/checking-symptom-checkers.

Case 7.2: Issues of Consent and Privacy in Establishing a Pregnancy Outcomes Registry

This case study was written by members of the case study author group.

Keywords Data protection, access and sharing; Privacy and confidentiality; Consent; Researcher roles and responsibilities

In the early months of the COVID-19 pandemic, maternal and foetal medicine specialists, obstetricians, midwives, general practitioners and pregnant women began to express concern about the impact that COVID-19 infection might have upon pregnancy outcomes and maternal health. Clinicians were limited by a lack of evidence in their ability to reassure women and support them to make informed decisions about their pregnancies and activities. Key questions included whether COVID-19 increased the risk of adverse outcomes such as miscarriage, stillbirth or premature labour; whether child development might be affected by COVID-19 infection at different points in a pregnancy; the relative safety of modes of delivery in the context of COVID-19 infection; whether it was safe to breastfeed when infected by COVID-19; and what impact courses of treatment might have upon foetal and maternal health.

Researchers and clinicians in a country in the global North established a registry to collect data about the outcomes of COVID-19 infection in pregnancy. Maternity carers across a range of health settings were asked to register and provide information on women suspected of having COVID-19 at any point during their pregnancy and up to 6 weeks post-partum. The original request to the national ethics committee for a waiver of consent was changed to a proposal for an opt-out approach. Clinicians registering women were asked to notify the women of their registration and provide them with details on how to be excluded from the registry if they wished. The plan was that data would be reported collectively and in a de-identified manner. In the early months of the registry, fewer than 10 cases were added, reflecting the effectiveness of the national COVID-19 response strategy. No women opted out, although it is possible that eligible women were not registered by their health provider. Without more cases, the researchers were concerned that analysing and reporting the data collected in the registry risked identifying the women involved.

In the context of the country's public health response, public health agencies are reporting "de-identified" information about COVID-19 cases in the community, including their area of domicile, age and sex, and details about their movements. This information is widely reported upon in the media.

Questions
1. In a pandemic should the same standards of de-identification of individual-level data apply when publishing public health information and research designed to inform clinical practice? Why?

2. How should the context of a pandemic affect the arguments for ensuring that sources of data cannot be identified?
3. Given the need to understand the clinical implications of COVID-19 for pregnancy and neonatal care, is there a case for an ethics committee reviewing this registry to approve a waiver of consent? Why?

Case 7.3: Ethical Conduct and Review of Research

This case study was written by members of the case study author group.

Keywords Ethical review; Privacy and confidentiality; Ethics committee remits and responsibilities; Consent; Qualitative research

Like some other Caribbean countries, little research is conducted in Country A. It is believed that this is due to the culture of the country, where executing research studies is considered atypical. Nonetheless, the importance of carrying out research studies has been recognized and is supported by local authorities. As a result, measures have been put into place to ensure that a national research ethics committee (NREC) exists to guide researchers and protect research participants. In addition, there is appreciation for the conduct of contextually relevant research studies whose findings can support decision-making as well as the development and reform of local policies. There are, however, some limitations, which make the use of an evidence-based approach challenging. Country A's small population size and the relatively low priority it gives to research are exacerbated by insufficient funding opportunities, limited research skills, inadequate ethics training and confidentiality issues – all of which often make it difficult for the full benefits of research to be realized.

During the COVID-19 pandemic, the possibility of conducting research in Country A was further compromised by the inability of the NREC to function optimally. Its members played a significant role in serving on the COVID-19 task force and were thus occupied with other priorities. The challenge of limited resources, including human resources, made it particularly difficult for the NREC to function as expected.

An early-career researcher from Country A submitted a research protocol to the NREC for review and approve shortly after the first COVID-19 case was reported in the country. The aim of the study was to assess the knowledge about, attitudes towards and practices regarding hand hygiene among health-care professionals who work at the country's main hospital. One of the objectives of this study was to obtain information on effective means of minimizing the spread of infectious diseases, like COVID-19. It was anticipated that the knowledge gained would guide the decision-makers, including the hospital administrator, to make recommendations that would promote more effective hand-hygiene practices, including more frequent hand-cleansing.

Although ethical review was expected to take a maximum of 6 weeks, as per the NREC's website, it took 12 weeks for the NREC to provide feedback on the protocol, which was not approved. The committee requested a number of changes prior to resubmission. The NREC considered that the methodology, which proposed using face-to-face interviews to obtain data, was not appropriate in this setting, which had a small number of specialized doctors, whose privacy and confidentiality would be compromised. It was particularly concerned about protecting the doctors from being easily identified, as naming a specialism was considered to equate to revealing the doctors' identities. The NREC also recommended that participants should not be asked to provide written consent, as this too would affect their privacy and compromise confidentiality. Its rationale was that this was minimal risk research so a formally documented signature was neither important nor required.

Questions
1. What kind of safeguards and methodological approaches are required to provide assurances that the confidentiality of a specific group of participants can be maintained? Does the COVID-19 pandemic make it harder than usual to adhere to these? Why?
2. Is it ethical to conduct research studies if the usual safeguards cannot be effectively implemented, especially during emergency situations like the COVID-19 pandemic? Why?
3. Given the significant pressures and workloads experienced by some members of ethics committees during the pandemic, what kind of support may be necessary and appropriate to enable effective and timely ethics review? How should capacity to conduct effective and timely ethical reviews be prioritized in pandemic responses?

Case 7.4: Informed Consent and Data Protection in the Context of Increased Use of Information and Communication Technologies

This case study was written by members of the case study author group.

Keywords Consent; Privacy and confidentiality; Data protection, access and sharing; Ethical review; Vulnerability and inclusion; Digital and remote healthcare and research; Qualitative research

The COVID-19 pandemic has prompted widespread use of technological tools, such as social media, to contact people who potentially meet the inclusion criteria for quantitative and qualitative research (Ploug and Holm 2015). Consent processes have also been adapted and modified in response to the pandemic. Before the pandemic, some predominantly qualitative investigations had already introduced online informed consent forms. As social distancing and restrictions on population movements were imposed during the pandemic, the use of online consent forms became increasingly common in qualitative studies and were also introduced in quantitative research (Gilbert et al. 2017). In addition, online platforms have increasingly been used to collect research data. Research ethics committees have had to consider the epidemiological, clinical, social and ethical implications of these changes in practice in the context of the pandemic, and to think about how research can be conducted online ethically and safely.

A research ethics committee in a research institution in Latin America received an application for a qualitative research project which aimed to learn about romantic attachment between members of a couple and emotional regulation difficulties in each member during the COVID-19 outbreak. The research sought to directly address the widespread impact of COVID-19 on mental health in communities experiencing lockdowns. According to the protocol, the respondent would be invited by email to participate in a survey about their experience of being confined with their partner for a long time. A link would appear in the invitation, which, when clicked, would display an informed consent form. The research was sponsored by a reputable university in the city, and this sponsorship was expected to generate trust and confidence in participants that the information they provide will be kept confidential. After they have consented, a new instruction would appear, which enabled participants to access the survey itself. The system would also ask the respondent to enter their email address. This step would be mandatory and ensure that participants would be able to receive the results of the survey.

The research ethics committee considered that it was important to check that the research fulfilled the core requirements for it to be considered ethical: research participants should be selected equitably, participants' privacy should be protected, and their data should be kept confidential. The committee noted that the use of an online consent form did not offer respondents the opportunity to have a conversation with the researcher in which they could ask any questions or voice any concerns about the research. In addition, the fact that only individuals who had access to information technologies could be surveyed would have an impact on the equitable selection of research participants.

The committee also considered that as this research dealt with intimate issues and collected very sensitive data through online platforms, it was crucial that participants could be guaranteed confidentiality. Online platforms for collecting research data from participants are potentially vulnerable to security breaches and have not proved to be absolutely reliable when protecting information. Participants' consent to share their life experiences would be given on the condition that they were not going to be identified through their answers. Researchers consequently had a responsibility to ensure the sensitive data of the participants were appropriately safeguarded as the data were collected, analysed and reported.

Questions
1. What ethical issues are most pertinent when seeking online consent to research?
2. When an online survey is conducted that requests intimate information from participants, what specific ethical challenges do researchers have a responsibility to address?
3. When reviewing research proposals that collect data online, what should research ethics committees pay special attention to?
4. Is it appropriate for participants to be recruited via email and required to provide their email address to take part in this study? Why?

References

Gilbert, M., A. Bonnell, J. Farrell, D. Haag, M. Bondyra, D. Unger, and E. Elliot. 2017. Click yes to consent: Acceptability of incorporating informed consent into an internet-based testing program for sexually transmitted and blood-borne infections. *International Journal of Medical Informatics* 105: 38–48. https://doi.org/10.1016/j.ijmedinf.2017.05.020.

Ploug, T., and S. Holm (2015). Routinisation of informed consent in online health care systems. *International Journal of Medical Informatics* 84(4): 229–236. https://doi.org/10.1016/j.ijmedinf.2015.01.003.

Case 7.5: Research into COVID-19 and Cancer in Populous Low-Income Neighbourhoods

This case study was written by members of the case study author group.

Keywords Vulnerability and inclusion; Data protection, access and sharing; Consent; Ethical review; Privacy and confidentiality; Boundaries between research, surveillance and clinical care; Non COVID-19 research

In early March 2020, the government in a Latin American country declared a health emergency because of the COVID-19 pandemic, and decreed a period of preventive and compulsory social isolation (PCSI) for the entire population. This meant the suspension of non-essential work and recreational activities and social gatherings of any kind. Remote education using online platforms was maintained, but participation was limited, especially in low-income populations, because of poor connectivity and poor access to the technology required.

The public health-care system was mainly focused on medical emergencies and caring for people with moderate to severe COVID-19. The Ministry of Health took steps to contain SARS-CoV-2 transmission in low-income populations, facilitating access to COVID-19 testing and care for COVID-19 patients. In coordination with university researchers, students and public health-care teams, door-to-door surveys were carried out in populous low-income neighbourhoods in order to find possible cases of COVID-19 and provide assistance with food and primary health care. In this context, valuable health and socio-economic data were collected about a wide range of people.

PCSI is an effective way of limiting infection, but people on low incomes have greater difficulties in complying with it. This is because of precarious and overcrowded housing, where strict isolation is not possible. Added to this, informal-sector work and poorly paid jobs, which in many cases were lost or suspended without financial compensation or support, generated worse economic conditions, which made it difficult for families to maintain proper hygiene and obtain the food and medicines they needed. All these social conditions generated a greater risk of infection by COVID-19 and neglect of pre-existing diseases in these populations.

Near the end of 2020, an institution conducting cancer research made a call for applications for funding aimed at investigations relating to cancer, COVID-19 and social vulnerability. Cancer is one of the main non-communicable chronic diseases in the country, resulting in high levels of mortality and morbidity. A research team with a background in cancer research and in conducting clinical studies submitted a proposal and was awarded a research grant. The clinical protocol for the study was presented to the ethics advisory committee of the university, which was the institution responsible for the study and for managing the grant.

A prevalence, observational and cross-sectional study of people with cancer and their eventual recovery from COVID-19 was proposed. The hypothesis of the study was that PCSI in contexts of socio-economic vulnerability causes people's health to deteriorate, worsening the progression of cancer and its associated comorbidities, and also generating a greater risk of SARS-CoV-2 infection. The pandemic also had

a negative impact on the public health system. Information about potential participants for the clinical study would be taken from the university's databases, which
contained the results of previous health surveys carried out by the research team. In
those surveys, people had been asked if they had chronic diseases, including cancer.

The study proposal included an analysis of epidemiological and clinical data.
Participants would be required to give written consent prior to inclusion in the study.
The study would use blood and urine sampling to determine participants' general
state of health. It would evaluate routine oncological serum markers and identify
potential new biomarkers, both for cancer and for COVID-19. The study also sought
to promote clinical care, by linking people with the health system and obtaining data
about the prevalence of cancer and associated comorbidities, and the epidemiology
of COVID-19 in these populations. In this sense, one of its most important aims was
to generate clinical and health-care recommendations for the health authorities, with
the intention of improving the quality of life in these populations.

However, when the clinical protocol of the study was presented to the committee,
they raised concerns about the proposed use of health data obtained from earlier
surveys to identify potential study participants. The committee stated that in order to
use the information from earlier surveys, people had to be informed about the
existence of the proposed study and about the future use of the data at the time the
earlier survey data were collected. Concerns were also raised about the recruitment
process, which would involve initial contact by telephone, followed by a visit to the
participant's home. The committee pointed out that an independent witness was
required for the consent process, as the surveyed candidates came from low-income
neighbourhoods and were considered a vulnerable population.

The researchers responded to the committee, stating that when the earlier surveys
were carried out, respondents were consulted about the possibility of using the
information provided to generate clinical and health-care recommendations. They
also noted that when the surveys were conducted, the current call for research
applications had not been issued, and this use of the data had not been planned. In
addition, the protocol had incorporated a proposal for an independent witness for the
consent process, so as to preserve the rights of vulnerable populations.

Questions
1. Should archived survey data be used to identify potential participants for a study
   which seeks to identify and address their health needs in the context of the
   COVID-19 pandemic, even if survey respondents have not consented to
   such use? Why?
2. How should this study, which seeks to analyse clinical and epidemiological data
   from potentially vulnerable populations where COVID-19's impact could exacerbate inequalities, be prioritized?
3. Should research with a direct therapeutic component be prioritized over
   non-therapeutic research in pandemics? Why?

References

Atan, Arif. 2020. Satu Malaysia sengsara, netizen kecam pesakit Covid-19 ke-136 "kuat merayap". *Malaysia Dateline*, April 8. https://malaysiadateline.com/satu-malaysia-sengsara-netizen-kecam-pesakit-covid-19-ke-136-kuat-merayap/.

Bambra, C., R. Riordan, J. Ford, and F. Matthews. 2020. The COVID-19 pandemic and health inequalities. *Journal of Epidemiology and Community Health* 74(11): 964–968.

Bayer, R., and A.L. Fairchild. 2000. Public health – Surveillance and privacy. *Science* 290(5498): 1898–1899.

Berg, Jessica, Laura D. Buccini, and Catherine Koepper. 2018. Informed consent for registries. In *Registries for evaluating patient outcomes: A user's guide*, ed. Richard E. Gliklich, N.A. Dreyer, and M.B. Leavy. Rockville: Agency for Healthcare Research and Quality. https://www.ncbi.nlm.nih.gov/books/NBK208622/.

Bhutta, Z.A. 2004. Beyond informed consent. *Bulletin of the World Health Organization* 82(10): 771–777.

Bull, S., P.Y. Cheah, S. Denny, I. Jao, V. Marsh, L. Merson, et al. 2015a. Best practices for ethical sharing of individual-level health research data from low- and middle-income settings. *Journal of Empirical Research on Human Research Ethics* 10(3): 302–313.

Bull, S., N. Roberts, and M. Parker. 2015b. Views of ethical best practices in sharing individual-level data from medical and public health research: A systematic scoping review. *Journal of Empirical Research on Human Research Ethics* 10(3): 225–238.

Cheah, P.Y., N. Jatupornpimol, B. Hanboonkunupakarn, N. Khirikoekkong, P. Jittamala, S. Pukrittayakamee, et al. 2018. Challenges arising when seeking broad consent for health research data sharing: A qualitative study of perspectives in Thailand. *BMC Medical Ethics* 19(1): 86.

CIOMS. 2016. *International ethical guidelines for biomedical research involving human subjects*. Geneva: Council for International Organizations of Medical Sciences.

EU General Data Protection Regulation (GDPR): Regulation (EU) 2016/679 of the European Parliament and of the Council of 27 April 2016 on the protection of natural persons with regard to the processing of personal data and on the free movement of such data, and repealing Directive 95/46/EC (General Data Protection Regulation), OJ 2016 L 119/1.

Fegan, G., P.Y. Cheah, and The Data Sharing Working Group. 2021. Solutions to COVID-19 data sharing. *Lancet Digital Health* 3(1): e6.

Gasser, U., M. Ienca, J. Scheibner, J. Sleigh, and E. Vayena. 2020. Digital tools against COVID-19: Taxonomy, ethical challenges, and navigation aid. *Lancet Digital Health* 2(8): e425–e434.

Ghafur, S., J. Van Dael, M. Leis, A. Darzi, and A. Sheikh. 2020. Public perceptions on data sharing: Key insights from the UK and the USA. *Lancet Digit Health* 2(9): e444–e446.

Global Research Collaboration for Infectious Disease Preparedness (GLOPID-R). 2018. *Principles of data sharing in public health emergencies*. https://www.glopid-r.org/wp-content/uploads/2022/07/glopid-r-principles-of-data-sharing-in-public-health-emergencies.pdf.

Han, Q., B. Zheng, M. Cristea, M. Agostini, J. Bélanger, B. Gützkow, et al. 2021. Trust in government regarding COVID-19 and its associations with preventive health behaviour and prosocial behaviour during the pandemic: A cross-sectional and longitudinal study. *Psychological Medicine*: 1–11. https://doi.org/10.1017/S0033291721001306.

Homolak, J., I. Kodvanj, and D. Virag. 2020. Preliminary analysis of COVID-19 academic information patterns: A call for open science in the times of closed borders. *Scientometrics* 124(3): 2687–2701.

Ienca, M., and E. Vayena. 2020. On the responsible use of digital data to tackle the COVID-19 pandemic. *Nature Medicine* 26(4): 463–464.

Moorthy, V., A.M. Henao Restrepo, M.P. Preziosi, and S. Swaminathan. 2020. Data sharing for novel coronavirus (COVID-19). *Bulletin of the World Health Organization* 98(3): 150.

National Commission. 1979. *The Belmont report: Ethical principles and guidelines for the protection of human subjects of research*. Bethesda: National Commission for the Protection of

Human Subjects of Biomedical and Behavioral Research. https://www.hhs.gov/ohrp/sites/default/files/the-belmont-report-508c_FINAL.pdf.

Newlands, G., C. Lutz, A. Tamò-Larrieux, E.F. Villaronga, R. Harasgama, and G. Scheitlin. 2020. Innovation under pressure: Implications for data privacy during the Covid-19 pandemic. *Big Data & Society.* https://doi.org/10.1177/2053951720976680.

Osterrieder, A., G. Cuman, W. Pan-Ngum, P.K. Cheah, P.-K. Cheah, P. Peerawaranun, et al. 2021. Economic and social impacts of COVID-19 and public health measures: Results from an anonymous online survey in Thailand, Malaysia, the UK, Italy and Slovenia. *BMJ Open* 11(7): e046863. https://doi.org/10.1136/bmjopen-2020-046863.

Pratt, B., and S. Bull. 2021. Equitable data sharing in epidemics and pandemics. *BMC Medical Ethics* 22(136). https://doi.org/10.1186/s12910-021-00701-8.

Rios, R.S., K.I. Zheng, and M.H. Zheng. 2020. Data sharing during COVID-19 pandemic: What to take away. *Expert Review of Gastroenterology and Hepatology* 14(12): 1125–1130.

Rothstein, M.A. 2010. Is deidentification sufficient to protect health privacy in research? *American Journal of Bioethics* 10(9): 3–11.

Schneiders, M.L., B. Naemiratch, P.K. Cheah, G. Cuman, T. Poomchaichote, S. Ruangkajorn, et al. 2022. The impact of COVID-19 non-pharmaceutical interventions on the lived experiences of people living in Thailand, Malaysia, Italy and the United Kingdom: A cross-country qualitative study. *PLoS One* 17(1): e0262421.

Serwadda, D., P. Ndebele, M.K. Grabowski, F. Bajunirwe, and R.K. Wanyenze. 2018. Open data sharing and the Global South – Who benefits? *Science* 359(6376): 642–643.

Sheehan, M. 2011. Can broad consent be informed consent? *Public Health Ethics* 4(3): 226–235.

Smith, Maxwell, and Ross Upshur. 2019. Pandemic disease, public health, and ethics. In *The Oxford handbook of public health ethics*, ed. Anna C. Mastroianni, Jeffrey P. Kahn, and Nancy E. Kass. Oxford: Oxford University Press.

Sweeney, L. 2000. *Simple demographics often identify people uniquely*, Data privacy working paper 3. Harvard University. https://dataprivacylab.org/projects/identifiability/paper1.pdf.

Tangcharoensathien, V., J. Boonperm, and P. Jongudomsuk. 2010. Sharing health data: Developing country perspectives. *Bulletin of the World Health Organization* 88(6): 468–469.

Tindana, P., S. Molyneux, S. Bull, and M. Parker. 2019. "It is an entrustment": Broad consent for genomic research and biobanks in sub-Saharan Africa. *Developing World Bioethics* 19(1): 9–17.

Upshur, R.E. 2002. Principles for the justification of public health intervention. *Canadian Journal of Public Health* 93: 101–103.

Vayena, E., and A. Blasimme. 2017. Biomedical big data: New models of control over access, use and governance. *Journal of Bioethical Inquiry* 14(4): 501–513.

Chapter 8
Dimensions of Vulnerability

Luciana Brito and Ilana Ambrogi

Abstract The COVID-19 pandemic has been a public health emergency on a global scale, impacting all nations and peoples. As previous health emergencies demonstrated, even when the infectious agent is nonselective, people and contexts are affected differently. Frequently these differences are not due to individual characteristics but to precarious contexts that became even less safe during emergencies, and exacerbate inequalities. An unknown disease that affects the world in a rapid manner brings many challenges. These range from an initial lack of knowledge about the biological effects of the viral infection and how to treat it, to its impacts on resources and economies. Inequitable COVID-19 vaccine distribution can be understood as a categorical example of how the pandemic has had different impacts on different countries and populations, and has exacerbated vulnerabilities. The importance of a comprehensive and considered account of vulnerability in research ethics has been discussed for decades, and this chapter provides an overview of the concept of vulnerability by outlining three dimensions of vulnerability discussed in the literature: the individual, the structural and the relational. These dimensions can overlap and intersect in dynamic and relational ways, especially during public health emergencies such as the COVID-19 pandemic, highlighting the importance of paying attention to vulnerability and inclusion in research, and to the development of protections that account for vulnerabilities in research. The cases presented in this chapter provide examples of how the COVID-19 pandemic exacerbates pre-existing vulnerabilities and show why it is important to reflect on this. Specifically, they prompt consideration of ethical issues associated with excluding populations such as pregnant women and people with disabilities from research, conducting research

L. Brito (✉)
Anis - Instituto de Bioética, Brasília, Brazil
e-mail: l.brito@anis.org.br

I. Ambrogi
Anis - Instituto de Bioética, Brasília, Brazil

Programa de Pós-graduação em Bioética Ética Aplicada e Saúde Coletiva – PPGBIOS/Fiocruz/ENSP, Rio de Janeiro, Brazil

S. Bull et al. (eds.), *Research Ethics in Epidemics and Pandemics: A Casebook*,
Public Health Ethics Analysis 8, https://doi.org/10.1007/978-3-031-41804-4_8

with psychiatric patients, and conducting research in impoverished settings with heighted food insecurity during the COVID-19 pandemic.

Keywords COVID-19 pandemic · Research ethics · Public health emergencies · Vulnerability and inclusion · Risk/benefit analysis · Social and scientific value · Safety and participant protection · Research design and adaption · Researcher roles and responsibilities · Privacy and confidentiality · Data protection, access and sharing · Boundaries between research, surveillance and clinical care · Digital and remote healthcare and research · Resource allocation · Community engagement and participatory processes

8.1   Introduction

The COVID-19 pandemic has been a public health emergency (PHE) on a global scale, impacting all nations and peoples. As is already known from previous health emergencies, even when the pathological agent is nonselective, different people and countries are affected differently. Frequently these differences are not due to individual characteristics but to precarious contexts that became even less safe than before (Khetan et al. 2022). An unknown disease that affects the world in a rapid manner brings many challenges. These range from an initial lack of knowledge about the biological effects of the viral infection and how to treat it, to its impacts on resources and economies. Inequitable COVID-19 vaccine distribution can be understood as a categorical example of how the pandemic has had different impacts on different countries and populations, and has exacerbated vulnerabilities (Acharya et al. 2021; VOA News 2021; Basak et al. 2022; Fisseha et al. 2021).

The importance of a comprehensive and considered account of vulnerability in research ethics has been discussed for decades (Luna 2009, 2019; Hurst 2008; Lange et al. 2013). In order to provide an overview of the concept of vulnerability, we will examine at least three dimensions of vulnerability discussed in the literature: the individual, the structural and the relational. We will also show how these dimensions can overlap and intersect in dynamic and relational ways, especially during PHEs, such as the COVID-19 pandemic (Luna 2009, 2019; The Lancet 2020). The cases presented in this chapter provide examples of how the COVID-19 pandemic exacerbates pre-existing vulnerabilities and show why it is important to reflect on this. Clearly, there is an ethical imperative for research data to reflect the needs of the most vulnerable populations.

Economic and political instability, along with insufficient medical and social protection resources, have been important drivers of the worsening inequities and inequalities seen during the COVID-19 pandemic (Rocha et al. 2021; Etienne 2022; Busso and Messina 2020). The widening of social, racial, gender and economic inequalities worsens existing vulnerabilities and creates new ones. In this context, ethical issues must be carefully considered when research is carried out with populations who face historical and socio-economic structural inequalities. These considerations also affect how we should think about research ethics during PHEs with populations in disadvantaged contexts. The exclusion of groups or individuals as research participants is an approach that in many instances leads to the worsening of vulnerabilities, and inclusion strategies should always be adopted.

8.2   Conceptualizations of Vulnerability

On the individual level, the idea of vulnerability is related to the postulation that every individual has fragilities requiring safeguards (Butler 2010, 2016). Precariousness is understood as a condition shared by all living beings, owing to our dependence on contexts and our interdependence with each other, which has been defined as the "inherent source of vulnerability" (Lange et al. 2013; Butler 2010). As such, precariousness, or our inherent vulnerability, can be seen as an equalizing generalized condition of human beings. Yet, in certain situations where populations face structural determinants of inequality, this omnipresent condition of fragility is experienced differently.

Case 8.1 prompts reflection on the exclusion of pregnant women from clinical trials for the COVID-19 vaccine. The discussion of vulnerability proves essential when thinking about research ethics in this situation for at least two reasons. Against a background of gender inequity and the historical relativization of women's autonomy during pregnancy, important questions arise about when pregnant women are not offered the opportunity to make an informed decision about participation in vaccine trials for an emerging disease. At the start of the COVID-19 pandemic pregnant women were deemed too vulnerable to participate in vaccine trials, though not vulnerable enough to the effects of the virus to receive the vaccine during initial vaccine rollouts.

Pregnant women should not be considered vulnerable simply by virtue of being pregnant (CIOMS 2016). The consequences of an initial categorical exclusion from vaccine research were dramatic for pregnant women in the most precarious contexts. COVID-19 was found to be associated with an increased risk of maternal morbidity and mortality (Metz et al. 2022). Many countries in the global South had rates of maternal mortality due to COVID-19 five to ten times higher than those in countries in the global North (PAHO and WHO 2021). Most of the pregnant women hospitalized with COVID-19 were unvaccinated (Engjom et al. 2022). The assumption that a category of people (pregnant women) should be excluded from research trials on the basis of pre-existing vulnerabilities becomes fundamentally flawed when it creates and worsens vulnerabilities. It can be argued that evidence suggests that this exclusion from COVID-19 vaccine trials led, at least in some part, to harm, not protection, for pregnant women and exemplifies how, in research, vulnerabilities should guide our focus, not avert it.

Vulnerability also has structural dimensions; that is, there are economic, social and cultural dynamics that create power differentials (Palk et al. 2020). Contexts of inequality and inequity produce conditions for deprivation and oppression of certain individuals, groups or populations (Butler 2016; Malmqvist 2017). Researchers and bioethicists commonly describe individuals or groups who face structural social and health inequalities as vulnerable, as seen in the Helsinki Declaration (World Medical Association 2013). Initial ideas of vulnerability were mostly categorical – e.g. pregnant individuals (Case 8.1), psychiatric populations (Case 8.2) or extremely poor people (Case 8.3). Although the categorical understanding of vulnerability has been an important step towards trying to solve issues of vulnerability in research, it carries limitations.

Cases 8.1–8.4 in this chapter illustrate how a lack of social protection and public services, impoverishment, social and gender inequalities, disabilities, racism and discrimination compound the many other factors that expose some populations to "layers of vulnerability". The concept of "layers of vulnerability", proposed by Florencia Luna (2009), provides a framework that helps us understand how the different dimensions of vulnerability are connected and how they interact with each other: "The metaphor of a layer gives the idea of something 'softer', something that may be multiple and different, and that may be removed layer by layer. It is not 'a solid and unique vulnerability' that exhausts the category; there might be different vulnerabilities, different layers operating" (Luna 2009, p. 128). Some layers of vulnerability may have a cascading effect, "which may create further layers of vulnerability or worsen existing ones" (Palk et al. 2020, p. 161). The idea of dimensions that intersect, interact, overlap and affect each other can also aid in the understanding of how these layers create contexts that make people vulnerable and points to the limitations of a categorical approach to vulnerability.

When analysing Cases 8.3 and 8.4, it is important to note that poverty and social exclusion have been defined not only as structural issues, but also as relational and fluid processes that are thought to be "driven by unequal power relationships interacting across an economic, political, social and cultural dimension and operating at different levels, including individual, household, group, community, country and global levels" (Ravinetto et al. 2013). These processes are considered to be important conditions that create vulnerabilities in medical research because they lead to health inequalities and diminished autonomy (Popay 2010). Thus, the relational aspect of vulnerabilities speaks to how these conditions interplay and interact in making someone vulnerable in a specific context. Research is inherently a context that impacts these dimensions, and can, therefore, add layers or trigger a cascade of vulnerabilities. The Nuffield Council on Bioethics and the Council for International Organizations of Medical Sciences (CIOMS) also remark on how vulnerabilities are dependent on the context (CIOMS 2016; Nuffield Council on Bioethics 2015, 2020). This indicates the complexity of vulnerability as far as research processes are concerned, even though its consideration is central to ethical practice.

Case 8.2 gives us the opportunity to reflect on research during the COVID-19 pandemic with groups that have specific needs. As this case shows, face-to-face contact can be essential when interacting with institutionalized or hospitalized psychiatric participants. At the same time, researchers need to be aware that some situations might expose participants to a harmful emerging virus. Researchers must be very careful when assessing situations in which limitations on autonomy are already imposed by structures of power, such as in a psychiatric unit or detention centre. These contexts can demonstrate how structural imposition of restrictions on autonomy can trigger a cascade of vulnerability. Adding a pandemic to this situation should draw scrutiny, as participants might feel inclined to participate in research because of a lack of other visitors or because the staff are paying them less attention. Here we can see how a PHE creates an urgent demand for research on the pandemic's impact on populations who already live in vulnerable contexts. This highlights the importance of considering how best to care for and address the urgent needs of populations living in residential facilities with restrictions on their freedoms, including prisons and other detention facilities.

The cases presented in this chapter aim to provoke reflection not only on concepts of vulnerability, but also on the necessity of thinking creatively and broadly about how to foster conditions that guarantee an inclusive approach in the research setting while promoting protection for research participants and their communities. In the course of a health emergency, the information produced by scientific inquiry becomes essential – especially as a way of dealing with the emergency itself. However, PHE situations are also known to exacerbate the conditions for increased vulnerabilities or to trigger their layers (Nuffield Council on Bioethics 2020; Kass et al. 2019; Chattu and Yaya 2020). Thus, combining an acknowledgement of vulnerability as an inherent possibility of the human condition (the individual dimension) with a context-sensitive analysis (structural and relational dimensions) can provide a way to reach a broader comprehension of all that is at play when thinking about research participation (Rogers et al. 2012). In particular, the provision of an effective approach to protecting participants during emergencies requires careful analysis of and attention to the individual, structural and relational dimensions of vulnerability.

Case 8.4 exemplifies the absence of people with disabilities from research during the pandemic. The WHO has notified some of the ways people with disabilities can be disproportionally impacted during the COVID-19 outbreak (WHO 2020). The exclusion of people with disabilities from research increases their vulnerabilities and leaves significant knowledge gaps regarding their health, needs and rights. In addition, it is important to recognize that COVID-19 can have long-term effects, including chronic health problems and increased susceptibility to disease. These also disproportionately impact those who live with multiple health disparities (Briggs and Vassall 2021). Disability-inclusive approaches to research should not be restricted to responses during a PHE; research practices should be recognized as making a fundamental contribution to understanding impacts and creating adequate and effective responses. These approaches should not only address the challenges and barriers people with disabilities face but also be undertaken in coordination with beneficiaries and existing sectors in civil society, and with health and social programmes (Banks et al. 2021).

8.3   Vulnerability and Inclusion in Research Ethics

Bioethics principles such as autonomy, beneficence and maleficence represent important ethical values; nevertheless, without careful analysis of the complexities of specific contexts which produce vulnerabilities, these values can become abstract and may result in the application of principles in ways that are not appropriately responsive to the context (Nuffield Council on Bioethics 2020). The recognition of different dimensions of vulnerability also means that understanding of this concept may not be assumed to be self-evident. From this perspective, research procedures may be considered ethically sound by research ethics committees, but perceived differently by a community or by grassroots organizations (Nuffield Council on Bioethics 2020).

Although the need to take vulnerability into account in research studies necessitates protection strategies, the idea of vulnerability has also been used to justify the categorical exclusion of individuals or of whole categories of people from research studies. The systematic exclusion of groups historically labelled as vulnerable, such as those living with food insecurity (Case 8.3), pregnant women (Case 8.1), people with disabilities (Case 8.4) or institutionalized people (Case 8.2) has led to a lack of imagination about who needs care, how to support autonomy and how to provide care for these groups (Dashraath et al. 2020; Stemple et al. 2016; Palk et al. 2020). Exclusion reinforces a lack of reflection and experience amongst researchers and research ethics committees (REC) on how to conduct ethical research with these groups (O'Mathúna and Siriwardhana 2017; Spiegel 2017). The consequences of categorical exclusions in the name of participant protection have deepened the marginalization of certain people and groups, as well as precluding them from enjoying the possible benefits of research participation and hindering scientific advancements of relevance to them.

It is important to consider how to promote the participation of people who have been historically excluded from studies, especially as the need for inclusive approaches is increasingly being recognized (Nuffield Council on Bioethics 2020). Guaranteeing conditions for inclusion and participation has been seen as a central aspect of ethical practice in research (Diniz 2019). Pregnant women, as seen in Case 8.1, are an example of a group that has been traditionally excluded from drug and vaccine trials (Beigi et al. 2021; Krubiner et al. 2021). The exclusion of pregnant women from vaccine trials during the COVID-19 pandemic and the Zika epidemic demonstrates the deficiencies of much decision-making undertaken far from people's realities, similarly to the negative consequences of categorical exclusions evidenced in the HIV crisis (MacQueen and Auerbach 2018; Treatment Action Group 2021; HIV Prevention Trials Network 2009; Schuklenk 2003).

There is growing evidence attesting to the impact that considering the perspectives and opinions of those in the field can have on creating more inclusive research practices (Diniz 2019; MacQueen and Auerbach 2018). Efforts are needed to create conditions for inclusion and for benefit-sharing processes compatible with participants' and communities' world views. In settings where individuals are impacted by political, social or economic inequalities, participant protection procedures should also involve an analysis of the pre-existing layers of vulnerability that can be activated or added to by research participation.

Case 8.3 draws attention to how research in settings where even the most basic needs are not being met creates ethical tensions between the participants' reality and standard research procedures, which require sensitive and contextually appropriate resolution. This is particularly important when planning or conducting research with populations that live with extreme vulnerabilities and when, as a result, the research might be one of the few opportunities for them to gain information or even professional attention. Case 8.3 also demonstrates how the pandemic disproportionally increased vulnerabilities in populations experiencing historical and structural social, racial, gendered and economic disadvantages. In addition to identifying the factors that influence the impacts of the pandemic, it is also important that research focuses on investigating how to build conditions that will help to develop community-centred solutions for preparedness and response in a PHE (Maxmen 2021).

One way of reimagining ethical participant protection procedures is to listen to and understand people's stories, biographies and needs, as is often undertaken in qualitative studies (Diniz 2019; Palk et al. 2020). These require not only knowledge of the local context by the researchers and RECs, but also imaginative and creative processes. Qualitative research has shown novel intersectional models for the inclusion of research participants, building of trust, and joint individual or community participation in research design and evaluation (Abramowitz et al. 2015; Pratt et al. 2020; Den Hollander et al. 2018). These approaches have strengthened inclusion and accountability for the community and brought recognition and validity to research protocols.

8.4 Vulnerability and the Promotion of Protection and Accountability

Community-based approaches play an important role in the identification of layers of vulnerability during the design of a research protocol or in the course of conducting the study (WHO 2016). A participatory approach requires research practices that acknowledge the imbalances of power and allow for attitudes and processes that address inequalities using combinations of different strategies (Lee et al. 2008; Diniz and Ambrogi 2017). That is, even the allocation of resources by the global North to the global South can be perceived as ethically questionable when the most urgent needs of communities and individuals are not addressed (Schuklenk 2014). As some layers of vulnerability are not readily evident and might not be identifiable to an outsider, there is an unjustifiable risk that research practices may reproduce or worsen inequalities if these vulnerabilities are not accounted for during research processes.

Consequently, protocols incorporating categorical approaches to exclusion criteria often insufficiently account for vulnerability and provide partial protection at best. In fact, these approaches can add vulnerabilities, limit understanding and narrow the reach of research benefits. As researchers, we find that these processes can also stifle our imagination and limit our ability to think about how to conduct research with populations in the most precarious contexts. There is no set number of layers of vulnerability that can justify categorical exclusions. Categorical exclusions do not afford protection. The issue of vulnerability in research should be addressed by creating ways of inclusion and protection that account for the layers of vulnerability.

Recognition of the circumstances of all those involved in a study can inform and expand understandings regarding research practices. That is not to say that the roles of researchers, participants and other individuals who form part of the research ecosystem should become blurred or confusing. Instead, it points to the need to recognize the value of a multidisciplinary, intersectional and representative approach, and to use imagination and creativity when thinking about appropriate research processes. This requires research practices that acknowledge the inherent imbalances of power and allow for attitudes and processes that seek equity and protection through inclusion.

There is no single answer to the question of how issues of vulnerability and inclusion should be addressed in research ethics. The commitment, however, is an ethical one – identifying and creating conditions that support inclusion and the development of protections that account for vulnerabilities, including in research responses to public health emergencies. For this, people in contexts that place them in the most precarious situations must be at the centre of the discussions about research ethics and global health governance. It is important to shift away from systems and structures that reproduce and perpetuate inequalities through fixed exclusionary approaches to vulnerability in order to create new methods for ensuring research accountability. Horizontal and long-lasting relationships with communities can amplify perspectives and decrease distances between researchers and potential participants and their communities. A commitment to decrease vulnerabilities, improve protection and enhance ethical practices entails developing approaches, alongside local groups and communities, to promote inclusion.

Case 8.1: Should Pregnant Women Be Included in COVID-19 Vaccine Trials?

This case study was written by members of the case study author group.

Keywords Vulnerability and inclusion; Risk/benefit analysis; Social and scientific value; Safety and participant protection; Placebo control; Vaccines

Health-care personnel are at high risk of contracting COVID-19, particularly those who work in intensive care units or other emergency settings. As the pandemic has progressed, it has become clear that although preventive measures such as physical distancing, universal masking, frequent hand-washing, efficient testing and contact tracing have important roles to play in reducing the spread of infection, they cannot entirely mitigate the spread of the disease nor end the pandemic. Therefore, mass vaccination, as a means of primary prevention, is widely seen as the most promising strategy for managing the pandemic (Society for Maternal-Fetal Medicine 2020).

According to the WHO report on gender equity in the health-care workforce, an analysis covering 104 countries showed that women make up around 70% of this workforce (Boniol et al. 2019). The US Centers for Disease Control and Prevention have stated that pregnancy greatly increases the risk of mortality from COVID-19 (Centers for Disease Control and Prevention 2020). Concerns about the effects of COVID-19 on pregnant women and newborns have placed them in a high-risk group. As data are gathered, there is evidence to show that pregnant women with COVID-19 have a higher chance of giving birth prematurely and that their children tend to be admitted to the neonatal unit (Allotey et al. 2020). Therefore pregnant women would likely benefit from receiving vaccines that minimize the likelihood of being infected with SARS-CoV-2, and the severity of any breakthrough infections which result in COVID-19.

A protocol for a Phase III placebo-controlled trial of an inactivated COVID-19 vaccine to be tested on health-care professionals was presented to a research ethics committee in a Latin American country. A key inclusion criterion was that that participant must be a health care-professional in direct contact with possible or confirmed COVID-19 cases. Important exclusion criteria for females were pregnancy (confirmed by beta-hCG testing), breastfeeding or intent to engage in sexual relations with reproductive intent in the 3 months following vaccination. The main reason for excluding pregnant women from most vaccine trials was to avoid risks to fetal health, as preclinical trials had not included tests of vaccine candidates in pregnant animals.

Questions
1. What are the main ethical considerations regarding inclusion or exclusion of pregnant women in vaccine trials during a pandemic?
2. Considering that health-care workers are at increased risk of contracting COVID-19, and that two-thirds of these workers are women, should COVID-19 vaccine trials include pregnant health-care workers? Why?

3. If pregnant women are included in COVID-19 vaccine trials, what levels of risk are acceptable and justified? What implications do requirements to manage and minimize risks appropriately have for the design and conduct of trials?

4. When a research ethics committee reviews a clinical trial for a new vaccine in an accelerated pandemic pathway, are there results from pre-clinical or clinical research (such as studies with pregnant animals) that require special consideration if pregnant women are to be recruited? Why?

References

Allotey, J., E. Stallings, M. Bonet, M. Yap, S. Chatterjee, T. Kew, et al. 2020. Clinical manifestations, risk factors, and maternal and perinatal outcomes of coronavirus disease 2019 in pregnancy: Living systematic review and meta-analysis. *British Medical Journal* 370: m3320.

Boniol. M., M. McIsaac, L. Xu, T. Wuliji, K. Diallo, and J. Campbell. 2019. *Gender equity in the health workforce: Analysis of 104 countries*. Geneva: World Health Organization.

Centers for Disease Control and Prevention. 2020. *Pregnancy, breastfeeding, and caring for newborns*, December 28. https://www.cdc.gov/coronavirus/2019-ncov/need-extra-precautions/pregnancy-breastfeeding.html.

Society for Maternal-Fetal Medicine. 2020. *SARS-CoV-2 vaccination in pregnancy*. Statement, December 1. https://s3.amazonaws.com/cdn.smfm.org/media/2591/SMFM_Vaccine_Statement_12-1-20_(final).pdf.

Case 8.2: Ethics and Research Policy in a Forensic Psychiatric Hospital

This case study was written by members of the case study author group.

Keywords Vulnerability and inclusion; Safety and participant protection; Research design and adaption; Researcher roles and responsibilities; Privacy and confidentiality; Data protection, access and sharing; Risk/benefit analysis; Boundaries between research, surveillance and clinical care; Non COVID-19 research; Digital and remote healthcare and research

Research in forensic psychiatry, the branch of psychiatry concerning the assessment and treatment of offenders with mental health problems in prisons and secure hospitals, is always a challenging subject, owing to the vulnerabilities of psychiatric patients and to broader concerns about public interest. The COVID-19 pandemic has produced ethical challenges for both forensic psychiatry practice and research. Forensic psychiatry research addresses our understanding of mental health disorders and the legal aspects of specific cases, focusing on the relationship between mental illness and criminality. In Europe, there are strict guidelines for the ethical conduct of research in forensic psychiatry hospitals, to ensure that the interests of participants are appropriately protected. Forensic psychiatry hospitals contain a closed community of patients and prisoners with mental health disorders, who take part in many assessment tests. Face-to-face interactions for assessment are important, and the pandemic has prompted consideration of the ethical and legal implications of implementing tele-psychiatry for such assessments.

The COVID-19 pandemic had immediate effects on the quality management and design of research protocols at a forensic psychiatry hospital in Europe, both in terms of timeframes and of results. During the pandemic, health-care and research staff also had less interest in and capacity to conduct research, because of infection-control restrictions and increased workloads. The physical distancing measures implemented to reduce the risk of SARS-CoV-2 transmission among staff and patients led to restrictions on visitors, in-person legal proceedings and movement around the hospital premises. In implementing these measures, health-care staff needed to address an ethical dilemma: should they develop a strict policy concerning the safety of the patients and prisoners in order to reduce SARS-CoV-2 transmission, with consequences for the psychopathology of the patients; or should they develop a policy approach with fewer restrictions but increased risk of SARS-CoV-2 transmission? This choice could also impact the validity of research. A stricter approach would limit face-to-face interactions during assessments and was more likely to alter patients' psychopathy, potentially biasing research results.

To minimize transmission risks, there was a move to implement a procedure to select patients potentially suitable for online psychotherapy and to provide psychotherapy and psychoanalysis remotely. This approach would enable some care to be provided for some of the patients at greater risk but raised ethical issues. The provision of remote psychotherapy can raise concerns about privacy, confidentiality and data security, and national data protection regulations had implications for when tele-psychiatry could be used. Furthermore, psychotherapy and psychoanalysis both have a strong ethical focus on preserving the patient's independence and confidentiality. Questions also arose about the effectiveness of remote psychotherapy and psychoanalysis as discussed in the example below.

Most of the follow-up or case report studies at the hospital relied on face-to-face contact with vulnerable patients, such as those with psychoses, schizophrenia or substance-abuse disorders. In one population study, patients with substance-abuse disorders would take part in a follow-up outpatient programme after their discharge from the hospital. The programme was considered to provide a direct benefit to participants, by providing on-going psychotherapy and support, and reducing the likelihood of further criminal action. Such follow-up studies would typically be conducted in confined settings, including ambulances. With infection-control and other measures to promote the safety and well-being of patients as their priority, researchers discussed how their study could continue during the pandemic with the ethics committee of the hospital. The committee decided that face-to-face contact with the outpatients in such confined settings was not acceptable during the pandemic. The researchers agreed but considered that the assessment and psychometric tests could only be effectively conducted in face-to-face interactions; the results of tele-psychiatry interactions were not considered sufficiently precise either for therapeutic or for research purposes. Consequently the researchers decided to explore the possibility of conducting the research with the local committee for substance-abuse disorders. A decision was taken to develop a special isolated outpatient zone in the hospital, where participants would be tested for COVID-19 and could then enter the main hospital building with appropriate security and infection-control measures.

Questions

1. How might the ethical issues presented by psychotherapeutic research with psychiatric patients change or be intensified by a pandemic?
2. What ethical challenges arise when seeking to protect the interests of small numbers of vulnerable participants during the COVID-19 pandemic while also conducting research addressing their needs?
3. How should the anticipated risks and benefits of using online or face-to-face approaches to follow up research in this context be evaluated and how should they inform assessments of the ethics of the research?

Case 8.3: Studying the Impact of COVID-19 on Vulnerable Populations

This case study was written by members of the case study author group.

Keywords Vulnerability and inclusion; Researcher roles and responsibilities; Resource allocation; Boundaries between research, surveillance and clinical care; Qualitative research; Digital and remote healthcare and research

In March 2020, an African country confirmed its first cases of COVID-19 and introduced measures to curb the spread of the virus. The initial control measures included promoting regular hand-washing and social distancing, and required that schools, religious institutions, offices and shops close, permitting only essential services to continue. A full lockdown was instated, with a ban on the use of all private and public transport, and a night curfew. From June 2020, most offices and shops were allowed to reopen, provided social distancing was observed and hand-washing facilities were available, and face coverings were required in public places. Private and public transport resumed with a limited number of persons per vehicle. Schools, religious institutions, sports facilities, arcades and places of entertainment remained closed until October 2020, then gradually reopened, and closed again from June until July 2021 during a second lockdown. Apart from examination-year students, who were allowed to return to school in October 2020, all other children remained at home between March 2020 and December 2021, and the night curfew remained in place.

The COVID-19 response has had a marked impact on the health and education of vulnerable populations, as well as on their economic situation and their psychosocial well-being (Kansiime et al. 2021). To understand the impact of COVID-19 and the associated public health response on vulnerable populations, researchers in the country collected information from mothers and children from impoverished backgrounds and fishing communities, families of children with disabilities, and young sex workers, all of whom were participating in other on-going research studies. Through phone interviews, the impact study, which started in June 2020, assesses participants' knowledge of and concerns about COVID-19, the impact of the COVID-19-related public health response on participants' daily lives, and suggested actions to reduce spread and support participants.

While planning and conducting the impact study, during the first months of the pandemic, the research team encountered several ethical issues. During the first lockdown period, on-going studies were paused by the national regulatory authorities and the review of new research protocols was delayed as organizations adjusted to working in lockdown. This meant that some studies where ancillary care was usually provided to participants during scheduled study visits were unavailable (Kapumba et al. 2021; Kapumba et al. 2022). As soon as the research team was able to start the phone follow-up and COVID-19 impact interviews in June 2020, it was extremely careful about how it presented what the study was doing, to avoid raising expectations or making promises that would make the participants feel let down. The researchers clearly explained the purpose of the study, clarified their institutional links and listened to any concerns, in order to maintain trust. Nevertheless, various problems brought up by participants caused ethical concerns and raised questions about previously clear boundaries between research activities and public health responses. For example, many participants had not earned any income at all during the first lockdown, and some had run out of food. Others were not able to obtain medication for chronic conditions and had started feeling ill but could not get to a health facility as there was a ban on public and private transport. Ethical issues also arose around access to care in cases where a participant presented with COVID-19 symptoms, as services were limited (Singh et al. 2020). Although the government distributed some food, most participants did not receive any. As the lockdown continued, multiple research teams had to decide what kind of support could be offered as part of their duty of care as a research institution. When study participants presented with emergency situations, the research teams responded by organizing modest food and medicine deliveries to participants' homes. In exceptional cases ambulance services were arranged for critically ill study participants.

Questions
1. What responsibilities do research teams have to respond to the pandemic-related emergency health and nutritional needs of participants?
2. Should researchers' responsibilities differ depending on whether or not participants' needs relate directly to the study question(s) and methods? Why?
3. In a pandemic should the responsibilities of researchers and public health responders be different from the boundaries observed in normal times? Why?
4. If you were a member of a research ethics committee responsible for reviewing this study, what would you request the study team to do in advance, by way of preparation?

References

Kansiime, M.K., J.A. Tambo, I. Mugambi, M. Bundi, A. Kara, and C. Owuor. 2021. COVID-19 implications on household income and food security in Kenya and Uganda: Findings from a rapid assessment. *World Development*, January 137:105199. https://doi.org/10.1016/j.worlddev.2020.105199.

Kapumba, B.M., N. Desmond, and J. Seeley. 2021. What do we know about ancillary care practices in East and Southern Africa? A systematic review and meta-synthesis. *Wellcome Open Research* 6:164. https://doi.org/10.12688/wellcomeopenres.16858.1.

Kapumba, B.M., N. Desmond, and J. Seeley. 2022. A chronological discourse analysis of ancillary care provision in guidance documents for research conduct in the global south. *BMC Medical Ethics* 23: 51. https://doi.org/10.1186/s12910-022-00789-6.

Singh, J.A., S.V. Bandewar, E.A. Bukusi. 2020. The impact of the COVID-19 pandemic response on other health research. *Bulletin of the World Health Organization*, September 1;98(9): 625–631. https://doi.org/10.2471/BLT.20.257485.

Case 8.4: Inclusion of Persons with Disabilities in COVID-19 Research

This case study was written by members of the case study author group.

Keywords Vulnerability and inclusion; Researcher roles and responsibilities; Social and scientific value; Community engagement and participatory processes; Resource allocation; Digital and remote healthcare and research

There is a high risk that persons with some types of disability are disproportionately impacted by COVID-19, for a variety of reasons. These include existing barriers to health services; disruption to health or rehabilitation services not related to COVID-19; the effects of containment and isolation on mental health; the implications of social distancing and restrictions on movement for people who require support from caregivers; and the disruption to disability-inclusive development programmes. All of these can have an adverse effect on the well-being, participation and economic security of persons with disabilities.

In a sub-Saharan African country, there was evidence that, directly or indirectly, COVID-19 has a more adverse effect on people with disabilities than on those without. An online survey was being conducted by university researchers in collaboration with a local non-governmental organization (NGO) to determine the impact of COVID-19 on access to routine health care among the general population. Participants were purposively recruited through the local NGO. During the process of data collection, the researchers realized that, for unknown reasons, persons with disabilities were not visible in the research. During the whole research journey – from the conceptualization of the idea and the development of the research team, to the partnership development, proposal development and ethics review application – persons with disabilities were not included. The next stages of the research were data analysis, writing up, dissemination, translation into practice and planning future strategies. The research team considered whether to take steps to include persons with disabilities at that stage, and if so, how. The team also wanted to consider how to maintain the quality of the research, ensure the data were comparable and adhere to the terms of research ethics approval. Ensuring that persons with disabilities are represented in such ventures is a matter for which there is advocacy in the country.

If the research team was to add persons with disabilities, this could compromise the coherence of the research and would require the researchers to submit a protocol amendment for review by the ethics committee. In particular, the description of the sample population in the original research proposal did not allow for the inclusion of an additional population identified after the study had started. If such a population were to be included it would require changes in the research to enable reasonable accommodations for people with disabilities, which were most likely not budgeted for. However, the researchers realized that omitting persons with disabilities would mean that the impact of the COVID-19 pandemic on such persons would not be tracked. This was likely to compromise the effectiveness of consequent health policy decisions by the government, because without these data they would not be well placed to introduce sound prevention and mitigation interventions for this part of the community. Also, if this group was omitted, it would have an impact on the future connections between the researchers and the excluded population.

Questions
1. Do researchers studying the impacts of the pandemic on access to routine health care have an obligation to ensure that all groups in society are represented in their findings, even if some are more difficult to reach than others? Why?
2. Should the research team ensure that persons with disabilities are included in the research? Why?
3. How could this research have been inclusive from its inception, respecting the vulnerability of disabled persons?
4. Should the research team support the rights of persons with disabilities to equal access to health care during the pandemic, or is this beyond their remit? Why?

References

Abramowitz, S.A., K.L. Bardosh, M. Leach, B. Hewlett, M. Nichter, and V.K. Nguyen. 2015. Social science intelligence in the global Ebola response. *The Lancet* 385(9965): 330.

Acharya, K.P., T.R. Ghimire, and S.H. Subramanya. 2021. Access to and equitable distribution of COVID-19 vaccine in low-income countries. *npj Vaccines* 6. https://doi.org/10.1038/s41541-021-00323-6.

Banks, L.M., C. Davey, T. Shakespeare, and H. Kuper. 2021. Disability-inclusive responses to COVID-19: Lessons learnt from research on social protection in low- and middle-income countries. *World Development* 137: 105178. https://doi.org/10.1016/j.worlddev.2020.105178.

Basak, P., T. Abir, A. Al Mamun, N.R. Zainol, M. Khanam, M.R. Haque, et al. 2022. A global study on the correlates of gross domestic product (GDP) and COVID-19 vaccine distribution. *Vaccines* 10(2): 266. https://doi.org/10.3390/vaccines10020266.

Beigi, R.H., C. Krubiner, D.J. Jamieson, A.D. Lyerly, B. Hughes, L. Riley, R. Faden, and R. Karron. 2021. The need for inclusion of pregnant women in COVID-19 vaccine trials. *Vaccine* 39(6): 868–870.

Briggs, A., and A. Vassall. 2021. Count the cost of disability caused by COVID-19. *Nature* 593: 502–5055. https://www.nature.com/articles/d41586-021-01392-2.

Busso, M., and J. Messina. 2020. *The inequality crisis: Latin America and the Caribbean at the crossroads*. Washington, DC: Inter-American Development Bank. https://publications.iadb.org/en/the-inequality-crisis-latin-america-and-the-caribbean-at-the-crossroads.

Butler, J. 2010. *Frames of war: When is life grievable?* London: Verso.

Butler, Judith. 2016. Rethinking vulnerability and resistance. In *Vulnerability in resistance*, ed. Judith Butler, Zeynep Gambetti, and Leticia Sabsay, 12–27. London: Duke University Press. https://doi.org/10.1515/9780822373490.

Chattu, V.K., and S. Yaya. 2020. Emerging infectious diseases and outbreaks: Implications for women's reproductive health and rights in resource-poor settings. *Reproductive Health* 17(1): 1–5.

CIOMS. 2016. *International ethical guidelines for health-related research involving humans*. 4th ed. Geneva: Council for International Organizations of Medical Sciences. https://cioms.ch/wp-content/uploads/2017/01/WEB-CIOMS-EthicalGuidelines.pdf.

Dashraath, P., K. Nielsen-Saines, S.A. Madhi, and D. Baud. 2020. COVID-19 vaccines and neglected pregnancy. *The Lancet* 396(10252): E22.

Den Hollander, G.C., J.L. Browne, D. Arhinful, R. Van Der Graaf, and K. Klipstein-Grobusch. 2018. Power difference and risk perception: Mapping vulnerability within the decision process of pregnant women towards clinical trial participation in an urban middle-income setting. *Developing World Bioethics* 18(2): 68–75. https://pubmed.ncbi.nlm.nih.gov/27761986/5.

Diniz, D. 2019. Enlarging bioethics imagination in humanitarian settings. *Developing World Bioethics* 19(3): 124–125.

Diniz, D., and I. Ambrogi. 2017. Research ethics and the Zika legacy in Brazil. *Developing World Bioethics* 17(3): 142–143.

Engjom, H., T. van den Akker, A. Aabakke, O. Ayras, K. Bloemenkamp, S. Donati, et al. 2022. Severe COVID-19 in pregnancy is almost exclusively limited to unvaccinated women – Time for policies to change. *Lancet Regional Health - Europe* 13: 100313. http://www.thelancet.com/article/S2666776222000060/fulltext.

Etienne, C.F. 2022. COVID-19 has revealed a pandemic of inequality. *Nature Medicine* 28. https://www.nature.com/articles/s41591-021-01596-z.

Fisseha, S., G. Sen, T.A. Ghebreyesus, W. Byanyima, D. Diniz, H.H. Fore, et al. 2021. COVID-19: The turning point for gender equality. *The Lancet* 398(10299): 471–474. https://doi.org/10.1016/S0140-6736(21)01651-2.

HIV Prevention Trials Network. 2009. *HIV Prevention Trials Network: Ethics guidance for research*. https://www.hptn.org/sites/default/files/inline-files/HPTNEthicsGuidanceV10Jun2009_0.pdf.

Hurst, S.A. 2008. Vulnerability in research and health care: Describing the elephant in the room? *Bioethics* 22: 191–202.

Kass, N., J. Kahn, A. Buckland, A. Paul, and The Expert Working Group. 2019. *Ethics guidance for the public health containment of serious infectious disease outbreaks in low income settings: Lessons from Ebola.* Baltimore: Johns Hopkins Berman Institute of Bioethics. https://bioethics.jhu.edu/wp-content/uploads/2019/03/Ethics20Guidance20for20Public20Health20Containment20Lessons20from20Ebola_April2019.pdf.

Khetan, A.K., S. Yusuf, P. Lopez-Jaramillo, A. Szuba, A. Orlandini, N. Mat-Nasir, et al. 2022. Variations in the financial impact of the COVID-19 pandemic across 5 continents: A cross-sectional, individual level analysis. *eClinicalMedicine* 44: 101284. http://www.thelancet.com/article/S2589537022000141/fulltext.

Krubiner, C.B., Faden, R.R., Karron, R.A., Little, M.O., Lyerly, A.D., Abramson, J.S., et al. 2021. Pregnant women & vaccines against emerging epidemic threats: Ethics guidance for preparedness, research, and response. *Vaccine* [Internet], January. [cited 2021 Nov 5];39(1):85–120. Available from: https://doi.org/10.1016/j.vaccine.2019.01.011.

Lange, M.M., W. Rogers, and S. Dodds. 2013. Vulnerability in research ethics: A way forward. *Bioethics* 27(6): 333–340.

Lee, C., W.A. Rogers, and A. Braunack-Mayer. 2008. Social justice and pandemic influenza planning: The role of communication strategies. *Public Health Ethics* 1(3): 223–234.

Luna, F. 2009. Elucidating the concept of vulnerability: Layers not labels. *International Journal of Feminist Approaches to Bioethics* 2(1): 121–139.

———. 2019. Identifying and evaluating layers of vulnerability – A way forward. *Developing World Bioethics* 19(2): 86–95.

MacQueen, K.M., and J.D. Auerbach. 2018. It is not just about "the trial": The critical role of effective engagement and participatory practices for moving the HIV research field forward. *Journal of the International AIDS Society* 21(Suppl 7): e25179. https://doi.org/10.1002/jia2.25179.

Malmqvist, E. 2017. Better to exploit than to neglect? International clinical research and the non-worseness claim. *Journal of Applied Philosophy* 34(4): 474–488.

Maxmen, A. 2021. Inequality's deadly toll. *Nature*, April 28. https://pulitzercenter.org/es/node/21925.

Metz, T.D., R.G. Clifton, B.L. Hughes, G.J. Sandoval, W.A. Grobman, G.R. Saade, et al. 2022. Association of SARS-CoV-2 infection with serious maternal morbidity and mortality from obstetric complications. *Journal of the American Medical Association* 327(8): 748–759. https://jamanetwork.com/journals/jama/fullarticle/2788985.

Nuffield Council on Bioethics. 2015. *Children and clinical research: Ethical issues.* London: Nuffield Council on Bioethics. http://nuffieldbioethics.org/project/children-research.

———. 2020. *Research in global health emergencies: Ethical issues.* London: Nuffield Council on Bioethics. https://www.nuffieldbioethics.org/publications/research-in-global-health-emergencies/.

O'Mathúna, D., and C. Siriwardhana. 2017. Research ethics and evidence for humanitarian health. *The Lancet* 390(10109): 2228–2229.

PAHO and WHO. 2021. *Epidemiological update: Coronavirus disease (COVID-19).* December 2. Washington, DC: Pan American Health Organization and World Health Organization. https://iris.paho.org/handle/10665.2/55322.

Palk, A.C., M. Bitta, E. Kamaara, D.J. Stein, and I. Singh. 2020. Investigating assumptions of vulnerability: A case study of the exclusion of psychiatric inpatients as participants in genetic research in low- and middle-income contexts. *Developing World Bioethics* 20(3): 157–166. https://pubmed.ncbi.nlm.nih.gov/31943750/.

Popay, J. 2010. Understanding and tackling social exclusion. *Journal of Research in Nursing* 15(4): 295–297.

Pratt, B., P.Y. Cheah, and V. Marsh. 2020. Solidarity and community engagement in global health research. *American Journal of Bioethics* 20(5): 43–56.

Ravinetto, R.M., et al. 2013. Rapid response: Re: Revising the Declaration of Helsinki. *British Medical Journal* 346. https://www.bmj.com/content/346/bmj.f2837/rr/650200.

Rocha, R., R. Atun, A. Massuda, B. Rache, P. Spinola, L. Nunes, et al. 2021. Effect of socioeconomic inequalities and vulnerabilities on health-system preparedness and response to COVID-19 in Brazil: A comprehensive analysis. *Lancet Global Health* 9(6): e782–e792. http://www.thelancet.com/article/S2214109X21000814/fulltext.

Rogers, W., C. Mackenzie, and S. Dodds. 2012. Why bioethics needs a concept of vulnerability. *International Journal of Feminist Approaches to Bioethics* 5(2): 11.

Schuklenk, U. 2003. AIDS: Bioethics and public policy. *New Review of Bioethics* 1(1): 127–144.

———. 2014. Bioethics and the Ebola outbreak in West Africa. *Developing World Bioethics* 14(3): ii–iii. https://onlinelibrary.wiley.com/doi/10.1111/dewb.12073.

Spiegel, P.B. 2017. The humanitarian system is not just broke, but broken: Recommendations for future humanitarian action. *The Lancet* 6736(17): 1–8. https://doi.org/10.1016/S0140-6736(17)31278-3.

Stemple, L., P. Karegeya, and S. Gruskin. 2016. Human rights, gender, and infectious disease: From HIV/AIDS to Ebola. *Human Rights Quarterly* 38(4): 993–1021. https://doi.org/10.1353/hrq.2016.0054.

The Lancet. 2020. Redefining vulnerability in the era of COVID-19. Editorial. *The Lancet* 395(10230): 1089. http://www.thelancet.com/article/S0140673620307571/fulltext.

Treatment Action Group. 2021. *HIV research advocacy.* https://www.treatmentactiongroup.org/wp-content/uploads/2021/10/TAG_HTVN_HIV_research_advocacy_final.pdf.

VOA News. 2021. WHO, UNICEF say 130 countries yet to administer any COVID-19 vaccine, February 10. https://www.voanews.com/covid-19-pandemic/who-unicef-say-130-countries-yet-administer-any-covid-19-vaccine.

WHO. 2016. *Good participatory practice guidelines for trials of emerging (and re-emerging) pathogens that are likely to cause severe outbreaks in the near future and for which few or no medical countermeasures exist (GPP-EP): Outcome document of the consultative process.* https://www.who.int/blueprint/what/norms-standards/GPP-EPP-December2016.pdf?ua=1.

———. 2020. *Disability considerations during the COVID-19 outbreak.* Geneva: World Health Organization. https://www.who.int/publications/i/item/WHO-2019-nCoV-Disability-2020-1.

World Medical Association. 2013. WMA Declaration of Helsinki – Ethical principles for medical research involving human subjects. https://www.wma.net/policies-post/wma-declaration-of-helsinki-ethical-principles-for-medical-research-involving-human-subjects/. Updated Oct 2013.

Chapter 9
Participant Recruitment, Consent and Post-trial Access to Interventions

Maru Mormina, Halina Suwalowska, and Mira L. Schneiders

Abstract Humanitarian emergencies, including public health crises such as epidemics, can overwhelm local resources and severely disrupt the functioning of communities and societies. Conducting research during or in the immediate aftermath of an emergency poses increased practical and ethical challenges, not least because the need to rapidly generate valuable knowledge must be constantly balanced with the principles of humanitarian assistance. This chapter provides an overview of key ethical considerations relevant to recruitment, consent and post-trial access to interventions in pandemic contexts, and proposes an "ethics in practice" approach. Research conducted during emergencies is unavoidably context – and time – sensitive, making generalized guidance difficult. The aim of this chapter is thus not to prescribe a checklist for decision-making, but to assist researchers and practitioners to reflect on and discern what constitutes ethical practice during exceptional times. In particular, public health emergencies highlight tensions that can arise between balancing the rights and interests of research participants with the health needs of the population. Careful consideration is also needed of the necessity of minimising risks and maximising benefits, including ensuring that recruitment processes are sensitive to potentially altered risk perceptions and impacts of increased vulnerability on power imbalances. The importance of establishing and maintaining trust is reviewed, particularly when asymmetries in knowledge and access to resources are heightened in complex and challenging pandemic contexts. The five case studies presented in this chapter invite readers to reflect on ethical challenges that research during public health emergencies presents, particularly in

Authors Maru Mormina and Halina Suwalowska have contributed equally to this work. Each has the right to list themselves as first author.

M. Mormina · H. Suwalowska
The Ethox Centre, Nuffield Department of Population, University of Oxford, Oxford, Oxfordshire, UK

M. L. Schneiders (✉)
Socio-Ecological Health Research Unit, Institute of Tropical Medicine, Antwerp, Belgium
e-mail: mschneiders@itg.be

© PAHO and Editors 2024

S. Bull et al. (eds.), *Research Ethics in Epidemics and Pandemics: A Casebook*,
Public Health Ethics Analysis 8, https://doi.org/10.1007/978-3-031-41804-4_9

connection with processes for communicating with and recruiting participants which have been adapted in pandemic contexts; potential risks to research participants and study staff; and with the rights participants in control groups may have to access experimental products.

Keywords COVID-19 pandemic · Research ethics · Public health emergencies · Consent · Vulnerability and inclusion · Privacy and confidentiality · Safety and participant protection · Researcher roles and responsibilities · Research design and adaption · Digital and remote healthcare and research · Risk/benefit analysis · Community engagement and participatory processes · Research priority setting · Access to experimental treatments · Resource allocation · Regulatory review · Vulnerability and inclusion · Post trial follow up and monitoring

9.1   Introduction

Humanitarian emergencies, including public health crises such as epidemics, are large-scale natural or human-induced phenomena that overwhelm local resources and severely disrupt the functioning of a community or society. Conducting research during or in the immediate aftermath of an emergency poses increased practical and ethical challenges, not least because the need to rapidly generate valuable knowledge must be constantly balanced with the principles of humanitarian assistance. The five case studies presented in this chapter invite readers to reflect on some of the ethical challenges that research during public health emergencies presents, particularly in connection with participant recruitment and consent and with the post-trial obligations of researchers to participants.

   In this introduction to this set of case studies, we highlight three key ethical considerations relevant to recruitment, consent and post-trial access to interventions. While our reflections are grounded in established principles that govern research ethics and are endorsed in key international guidelines, their application will depend on the unique set of circumstances in which activities take place. We thus suggest an "ethics in practice" approach. Research conducted during emergencies is unavoidably context- and time-sensitive, making generalized guidance difficult. Our aim here is not to prescribe a checklist for decision-making but to assist researchers and practitioners to reflect on and discern what constitutes ethical practice during exceptional times.

9.2   Balancing the Rights and Interests of Research Participants with the Health Needs of the Population

Public health emergencies, such as the COVID-19 pandemic, acutely highlight some of the tensions that can arise when trying to balance the interests of individuals (e.g. research participants and/or patients) with those of the population as a whole.

To this end, it may be helpful to consider the key ethical principles and values that underpin medical and research ethics, in order to weigh up ethical considerations when there is a conflict between the interests of an individual and the wider population.

In the context of medical practice, clinical or medical ethics are primarily concerned with navigating the conflicts arising when determining what's best for the specific patient (Wilkinson et al. 2008). Research ethics, on the other hand, addresses conflicts that arise when balancing the interests of research participants/patients in general against those of medical science (CIOMS 2016). In research, as well as in clinical practice, respect for the participant's or patient's autonomy is a central ethical concern, highlighted by the emphasis on individual informed consent (to undergo treatment or to participate in research). In the context of a pandemic, in which the whole population is significantly impacted by an infectious disease outbreak, and collective action is fundamental to controlling it, individual-level concerns must be weighed against concerns for the population as a whole (population health ethics). This notwithstanding, it is essential that the need to reduce suffering for the population as a whole is carefully balanced against the rights and interests of individuals (Nuffield Council on Bioethics 2020).

One area where this need for balance is particularly evident is in the consent process. Seeking informed and voluntary consent is a fundamental way of ensuring that the rights and best interests of the individual are respected. However, where conditions on the ground are such that standard consent procedures become unfeasible, other appropriate approaches may need to be developed. In these cases, the principle of minimizing risks to health (such as the risks arising from face-to-face consent procedures) needs to be balanced against the principle of respect for individuals, which requires participation in research to be voluntary, informed and competent (CIOMS 2016; Nuffield Council on Bioethics 2020). Such conditions may be more difficult to determine when the interaction between researchers and participants is physically distanced and takes place remotely. Balancing these two principles may require that in addition to developing adaptive and innovative consent procedures that are ethically sound and work in an emergency context (e.g. remote consent via telephone), researchers need to pay particular attention to the relational aspect of the consent process (Nuffield Council on Bioethics 2020; WHO 2020).

The need to adapt consent processes in emergency contexts can, in fact, give rise to ethically problematic situations, for example with regard to assessing the validity and voluntariness of consent taken over the phone, especially when this involves minors, as discussed in Case 9.1. While these are practical difficulties for which there are no perfect solutions, allowing time to establish rapport with participants and creating a safe space of mutual recognition and respect will play an important role. Furthermore, the extent to which consent should be viewed as an on-going and dynamic process, rather than a discrete event, requires consideration. Case 9.2 offers an illustrative example of the need for such dynamic consent, especially when rapidly changing situations can lead to varying attitudes, behaviours and practices.

During public health emergencies such as the COVID-19 pandemic, unusual circumstances may arise where obtaining individual-level consent is not practical or even possible at all (Goldman and Gelinas 2021; Goyal et al. 2021; Largent et al. 2021). Case 9.3 prompts us to reflect on whether research ethics committees should make exceptions to standard requirements for individual informed consent in the context of COVID-19, and if so what the limits to this should be. Similarly, Case 9.4 asks us to consider how – in the context of a pandemic – consent processes for post-mortem research should be adapted and whether family consent can be ethically waived, given the heightened infection risks associated with consent protocols involving face-to-face and group consultations. In these and other cases, such as research conducted in emergency situations with patients who are comatose or incompetent, waiving consent may be justifiable if the research can demonstrate high social and scientific value and the waiver of consent is accompanied by community-level consultations to ensure acceptability (CIOMS 2016; Largent et al. 2021). In deciding to adapt or waive consent procedures, it is important to consider not only risk factors (see also below), cultural considerations and logistics, but also whether a waiver can be justified by the potential benefits that the research may generate.

Balancing population health needs against the needs and rights of participants requires going beyond consent, and including considerations about post-research obligations and the fair sharing of benefits (CIOMS 2016). This may require reassessing what is owed to participants/patients after the completion of a research study or trial (e.g. sharing research results, data and/or knowledge and interventions generated by the research, building local health infrastructure and capacity) (Lairumbi et al. 2011, 2012). For example, Case 9.5 prompts us to reflect on whether participants assigned to the placebo arm of a clinical trial should be prioritized to receive the treatment over other priority groups, once its efficacy and safety have been established.

While established ethical principles require researchers to provide fair benefits to participants, these considerations need to be weighed against the broader public health needs at the time, and against considerations about fair allocation of resources, including health workers' time. Case 9.2 reminds us that in the midst of a health emergency, frontline staff can very quickly become overstretched, and therefore it is more important than ever to weigh up the social value of research against pressing public health needs. These value assessments will certainly have different weightings than if made in "normal" times and may result in a greater prioritization of frontline activity, especially when research may be important and of benefit to individuals and communities in the long term but does not directly contribute to the immediate emergency response.

Balancing the need to provide benefits to participants against broader public health needs may also entail adopting a broad view of research benefits, which includes not only short-term and direct benefits (e.g. priority access to drugs or effective treatments) but also long-term and indirect benefits. Research (including research conducted during health crises) may be justified in terms of the benefits that individual participants derive from improvements to overall population health (e.g. reduced transmission of infection as a result of effective treatments/vaccines). During outbreaks, heightened infection risk in the population represents a threat to every person, and so benefits at the population level (including reducing transmission and developing effective treatments and vaccines) are also likely to translate into benefits at the individual level (including a reduced risk of infection and a higher chance of survival). However even when no direct benefits of research participation are envisaged, researchers still have a duty of care towards participants and should minimize risks and address any adverse effects resulting from the research, as we outline below. It is also important that questions about post-research access to interventions and obligations to research participants are carefully considered prior to the enrolment of participants so that these can be clearly communicated from the outset. This will help ensure that participants can weigh up the benefits and burdens of the proposed study and come to an informed decision about whether to take part.

9.3 Minimizing Risks and Maximizing Benefits

Infectious disease outbreaks are challenging circumstances in which to conduct health research. While most of the ethical issues that emerge when carrying out research during epidemics are similar to those encountered in "normal" times, one key difference involves the assessment of risk during an epidemic or pandemic. During outbreaks, risks inherent in the research process may be faced by researchers and frontline staff, as well as participants. There is a duty of care to consider the well-being of participants alongside the welfare of frontline staff, including researchers. Therefore, actions must be taken to identify and mitigate foreseeable risks generated by epidemics and pandemics (see Cases 9.2 and 9.4) (Nuffield Council on Bioethics 2020).

Under normal circumstances, health research entails both risks and benefits to participants, and for this reason, any potential physical, psychological or social harms must be identified prior to recruiting participants and justified in terms of the scientific and social value of the research (CIOMS 2016; Nuffield Council on Bioethics 2020). The value of research is understood as generating knowledge to reduce suffering and improve people's health. During emergencies, it may be particularly important to ensure evaluations of risks, benefits and the social and scientific value of research are informed by consultation with relevant populations. Measures to minimize and manage potential risks should be established, and a threshold of risk should be agreed upon, prior to ethical review. Participants must understand the risks of participating in the research and the measures being taken to manage such risks. However, during emerging infectious disease outbreaks, risks may be difficult to predict and evaluate, and researchers, participants and ethics committees have to make decisions in the context of uncertainty (Hofmann 2020).

During infectious disease outbreaks, existing vulnerabilities are often exacerbated and new ones can emerge, as discussed in Chap. 8. Under conditions of high vulnerability and uncertainty, participants may be inclined to make risk–benefit assessments that lead them to accept levels of risk they would not accept under normal circumstances (Macioce 2021). Recruitment processes must therefore be sensitive to risk perceptions and power imbalances that may result from increased vulnerability, as well as the social dynamics of the communities participating in the research (CIOMS 2016; Nuffield Council on Bioethics 2020). However, care should be taken that protection of vulnerable individuals or groups does not translate into recruitment criteria that unfairly disadvantage these groups, excluding them from the benefits of participating in research. Risk–benefit evaluations during research design, review and conduct need to include not only the risks associated with research participation but also any risks of non-participation, especially when a dearth of research may result, for example, in inadequate care for the groups excluded (Nuffield Council on Bioethics 2020).

9.4 Establishing and Maintaining Trust

Trust plays an important role in the conduct of scientific research. Yet trust is a complex concept, with no agreed definition, and therefore differences – even subtle ones – in understandings can lead to contradictory or inconsistent action, which can undermine the research enterprise (Resnik 2011). Trust is above all a relationship – between people, between people and groups, or between groups. It facilitates social cooperation (Whitbeck 1995), but it is not without cost: trust requires taking risks and embracing uncertainty (because there's no guarantee that the person or institution deemed trustworthy will act as expected). Trust can create moral and/or legal duties for the person/institution being trusted, whether implicitly or explicitly, to "keep their promise".

Because scientific research is a relational activity involving many collaborative interactions – between scientists, between scientists and research participants, between scientists and granting agencies, publishers, universities, etc. – trust is key (Resnik 2011; Kerasidou 2017). If parties do not fulfil the expectations placed upon them, the cooperation upon which much of the research enterprise depends becomes difficult, if not impossible. For this reason, trust goes hand in hand with respect. In the context of the researcher–participant relationship, trust is presupposed in the latter's consent to take part in the research (O'Neill 2002). Such presupposed trust in research places obligations on researchers to respect participants, including to avoid harming them, to protect their data, ensure confidentiality, act in the interests of society and not be unduly influenced by financial or other personal interests (Kass et al. 1996; Miller and Weijer 2006). Breaches of these obligations undermine trust and can lead to participants withdrawing from the research, or worse, declining to participate altogether. Case 9.4 demonstrates the importance of diligence and socio-cultural sensitivity when communicating with prospective participants and communities in order to establish and maintain trust. However, trust and corresponding obligations exist within highly asymmetrical relationships, owing to the power differential between researchers and participants (Karnieli-Miller et al. 2009), which renders the latter vulnerable and dependent on the former. This is especially so when the lines between research and intervention become blurred (Case 9.3), as is often the case in emergency contexts.

Crisis contexts such as pandemics and other public health emergencies can exacerbate asymmetries in knowledge and access to resources, and therefore power differentials, thus affecting perceptions of trustworthiness (Fox et al. 2020). For example, as mentioned above, heightened health needs may create false expectations about the benefits of research and could lead to misunderstandings, especially when research participation is perceived as the only way for individuals to satisfy basic needs (Kingori 2015). Such "empty choice" situations could be avoided, for example by collaborating with humanitarian or governmental organizations to provide services. Establishing, where possible, viable and visible referral mechanisms to such organizations, irrespective of research participation, may discourage enrolment in research studies as a measure of last resort to access basic services. Researchers also need to prevent misplaced trust and manage unrealistic expectations by being clear and transparent with participants and communities during the consent process about their research goals, including any limitations on post-trial access to experimental treatments, as illustrated by Case 9.5.

Trusting relationships between researchers, participants and local communities can be strengthened in emergency situations. Case 9.2 illustrates that trust in those tasked with representing the interests of the communities can sometimes offer legitimacy to a research study which is well beyond that conferred by the contractual model of informed consent (Appiah 2021). The relationship between trust, authority and consent deserves more attention than we can give it here. Suffice to say that, if research is a relational activity, as suggested above, the process of consent must also be relational, with trust as its implied currency. Researchers and the institutions they represent must trade in this currency with both individuals and communities. While emergency conditions may present additional challenges, appropriate engagement provides a forum for debate and negotiation that builds trust and safeguards the values and interests of the community as a whole at a time when this is most needed.

Nevertheless, trusting relationships can also be undermined by fear and misinformation. Uncertainty during crises alters normal patterns of information-seeking behaviour and this creates conditions rife for the spread of fake news and conspiracy theories, especially via social media (Huang and Carley 2020). The extent to which this can impact on the participant–researcher relationship is captured in Case 9.4, where challenges related to cultural acceptability are compounded by fears fuelled by misinformation. In complex environments like this, researchers have a greater responsibility to build trusting relationships with participants through effective engagement that enables participants to understand the acceptability of the proposed research, to respond to concerns and to ensure transparency and accountability. Equally critical is developing strong communication processes during recruitment and throughout the research to give participants clear information about the nature of the research and the potential benefits and risks, as well as addressing any conflicts of interest. Case 9.2 shows how people's attitudes to risk in rapidly evolving situations may be constantly changing as information (and misinformation) flows help consolidate new – albeit not always appropriate – risk management strategies, underscoring the importance of engagement and effective communication. Trustworthiness, together with timely, honest and effective communication, plays a key role in countering the diffusion of misinformation and promoting participation in research.

Conducting research in settings affected by public health emergencies or other humanitarian crises is challenging on many fronts, not least when it comes to accessing and recruiting participants. When face-to-face interactions pose significant personal and public health risks, online technologies can be used to seek consent (Case 9.4) or conduct interviews (Case 9.1). However, while online tools may offer the flexibility to circumvent participants' constraints and avoid health risks to researchers (Case 9.2), they also present ethical and methodological challenges (Brown 2018; Chiumento et al. 2018; Lo Iacono et al. 2016). The physical separation between researcher and participant can affect the research interaction itself, whether this is because of technical issues (connectivity, digital literacy) or simply due to the difficulties inherent in establishing a rapport and trust at a distance. Privacy and confidentiality may be more difficult to guarantee, especially if participants do not have access to private spaces or need to use shared devices (Case 9.1). Physical distance also increases the risk of misreading visual clues that may signal unease or lack of understanding. Finally, the disembodiment of the interaction may make it easier for participants to adopt a different persona or even fake their identity. All this can undermine trust and the authenticity of the relationship. When adapting research to online platforms, careful consideration is needed of researchers' responsibilities to get to know their participants, including allowing sufficient time for the interaction and being sensitive to unusual patterns of behaviour, especially when open exchange is not possible for whatever reason.

9.5   Conclusion

In sum, humanitarian emergencies such as infectious disease outbreaks require more, not less, ethical consideration, when difficult trade-offs between the common good and the good of the individual must be made; when an environment of heightened risk alters the balance between harms and benefits; and when trust can be strengthened by the shared struggle but also jeopardized by distorted perceptions and misinformation. Faced with decisions of life and death, researchers may feel that the moral imperative is to generate knowledge to minimize suffering, promote health and save lives, but this must not come at the cost of violating the paramount principle of respect for participants and communities. For those conducting research under these difficult circumstances, this entails a greater responsibility to develop ethically adaptive, responsive, innovative and proportionate approaches to research. Finding effective and efficient solutions under time pressure must be undertaken without neglecting compliance with internationally agreed ethical principles and guidelines. We recognize that achieving this requires ethical judgement to determine how such principles and guidelines should be applied in unique research contexts. We hope the cases in this chapter encourage ethical reflection and help researchers and reviewers who are confronted with the challenges of developing, reviewing and conducting research during public health emergencies.

Case 9.1: Ethical Challenges Arising When Recruiting Adolescent Minors by Telephone

This case study was written by members of the case study author group.

Keywords Consent; Vulnerability and inclusion; Privacy and confidentiality; Safety and participant protection; Researcher roles and responsibilities; Research design and adaption; Qualitative research; Non COVID-19 research; Digital and remote healthcare and research

This case describes the ethical challenges arising when informed consent for participation in research is sought by telephone from a minor. "John", aged 12, was living with his mother and members of his extended family. He shared a mobile phone with his mother and siblings. Researchers at a research institute in Africa were conducting research on the impact of HIV interventions among young people 12–17 years old. The researchers advertised the study on the community radio station, and used local networks to recruit the young people. When the country went into lockdown in response to COVID-19, the decision was made that all recruitment at the research institute had to be virtual. The researchers called John's mother to explain the study objectives and procedures, then asked if they could talk to John. John's mother expressed interest and was keen for John to participate in the study. The researchers cautioned John's mother that he had the right to refuse to participate even if she consented to him taking part. John's mother handed the phone to him as the family members shared the phone. John assented to take part in research.

John was reticent when discussing the research over the phone, and the researcher frequently offered him opportunities to ask questions or say if he was unclear about any part of the study. The researcher would often ask John to paraphrase the information provided about the study and in response he expressed the desire to proceed with the consent process without referring to the information. The researcher did not want to make John uncomfortable and so did not press him to repeat the study information. However, there were concerns about John's comprehension and whether he felt under pressure to hurry through the informed consent process. The researcher also wanted to ensure that John understood the consent process without keeping him on the phone for "too long". Consequently, the researcher was anxious about whether the consent was informed and voluntary, and whether to discontinue the informed process despite John's expressed desire to continue.

When John was asked if he could find a quiet private place to minimize disruptions and ensure his privacy during the call, he consulted his mother, and they decided together that he should remain in the same room as her. Given John's age and the socio-cultural context, the mother's concerns and protectiveness were understandable and respected. Still, the researcher wondered if John's privacy was compromised, and whether his reticence might have been caused by the lack of privacy.

Questions
1. Given the stigma associated with HIV in the community, the participant's lack of private telephone and private room, is it ethically acceptable to ask for consent to this type of research via phone? Why?
2. Given the researcher's concerns during the consent process, should John be enrolled in this study? Why?
3. How should the rights of adolescents be protected when consent processes must be conducted by phone?
4. What responsibilities do researchers have to ensure that consent to research sought over the phone is meaningful and voluntary?

Case 9.2: Quantitative and Qualitative Research into Attitudes Towards COVID-19

This case study was written by members of the case study author group.

Keywords Consent; Research design and adaption; Safety and participant protection; Risk/benefit analysis; Community engagement and participatory processes; Research priority setting; Qualitative research; Researcher safety; Digital and remote healthcare and research

At the onset of the COVID-19 pandemic, a team of social scientists set out to undertake a quantitative survey to determine knowledge, understanding and attitudes towards COVID-19 among residents of two large rural communities in an Asian country. Community leaders were approached and appraised of the research study, and their support for the study was obtained. It was expected that the results from the survey would help in the development of informational and educational strategies and guide public policy. A questionnaire was developed and applied using a cluster sampling technique. In total, 30 clusters (wherein all houses within a cluster would be surveyed) in each of the two rural communities were selected and the survey commenced. During the data collection process the researchers were assisted by community health workers who had received online training in administering the survey. The health workers were preoccupied with other responsibilities and, when seeking to administer the questionnaire, found that many prospective participants were reluctant to participate and needed encouragement to do so. Obtaining written consent was not always easy as respondents – who were largely non-literate – were disinclined to have the information sheet read out to them and to give their signature/thumb impression. They believed that as their community leaders were cognizant of this study and had given support, it was all right for them to participate.

Towards the end of the third month of data collection, the research team noticed a distinct reduction in people's sense of fear towards COVID-19. This was in sharp contrast to their experiences at the beginning of the process, when people feared even stepping out of their homes and were adhering to preventive measures such as face masks and social distancing. The health workers reported that in addition to a sense of complacency, people were resorting to the use of various traditional practices like consuming turmeric water and other herbal remedies that they felt would protect them from COVID-19.

To better understand this change in behaviour and corroborate the quantitative survey findings, the researchers sought to conduct a small qualitative study with a sample of residents from the two rural communities. One of the strengths of qualitative research is its capacity to explore and obtain deep insights into a phenomenon of interest. A revised proposal, with the proposed inclusion of a qualitative sub-study, was submitted to the research ethics committee for approval.

The qualitative study proposed to collect data via in-depth face-to-face interviews with participants. Each interview would take about an hour to complete as it would be an open-ended, exploratory process allowing respondents to express their thoughts, opinions and feelings. In addition, the interviews would need to be conducted privately and audio-recorded in a quiet place with minimal external noise. Invariably this would be inside the homes of the respondents, which are usually small and not always well ventilated. The risk of exposure to COVID-19 for both interviewer and interviewee was therefore higher than for the survey and it therefore did not receive ethics approval. One suggestion was to adapt the research methods and conduct the interviews via an online platform.

Questions
1. During the quantitative survey, researchers were required to seek individual written consent from participants. Should the requirement for such consent be revised in a pandemic context? Why?
2. When reviewing the proposed qualitative study, what considerations should the ethics committee take into account when determining whether the research risks are justifiable, managed and minimized to acceptable levels?
3. In this rural setting, some participants may be unable to participate in an online interview because of lack of access to the equipment needed, poor internet connectivity, and limited experience with online platforms. Given the need to minimize infection risks, what responsibilities arise to ensure that they do not get left out of research?
4. In this pandemic context, where the health workers assisting researchers already have multiple competing responsibilities, how should the proposed qualitative study be prioritized among other research priorities and public health priorities? Who should make such decisions and what ethical considerations should they take into account?

Case 9.3: Seeking Consent to Research Involving the Use of Convalescent Plasma from COVID-19 Donors in the Treatment of Cancer Patients

This case study was written by members of the case study author group.

Keywords Consent; Risk/benefit analysis; Researcher roles and responsibilities; Digital and remote healthcare and research

Cancer patients are at greater risk of developing COVID-19 than the general population, and those who contract it are at much greater risk of serious complications and death, probably as a result of the immunosuppression associated with their cancer treatment. In addition, cancer patients have a higher risk of developing blood clots, which can play a fundamental role in the pathogenesis of severe COVID-19 (Mei and Hu 2020).

The use of convalescent plasma from patients who have recovered from known viral infections has shown successful results in the treatment of SARS-CoV-1, MERS, influenza AH1N1 and Ebola, among others (Mupapa et al. 1999; Mair-Jenkins et al. 2015). In Latin America, a study was therefore proposed to explore the value of convalescent plasma in the population and specifically in adult cancer patients (aged 18+) who met criteria for severe COVID-19 with two or more poor prognostic factors for cancer. Patients would be grouped according to whether they had severe COVID-19 or COVID-19 with poor prognostic factors. These groups would then be subdivided into patients with and without cancer.

Following the administration of convalescent plasma to SARS-CoV-1 patients, varying results were reported, possibly attributable to the timing of the infusion (before or after 14 days from symptom onset) and lack of standardization of antibody titres. This study proposal involved the administration of two units of 200 ml of convalescent plasma to all patients who met the inclusion criteria and consented to take part, and analysis of the safety and impact of convalescent plasma on morbidity and mortality. The study would also evaluate the presence of other factors that affected patient outcomes in terms of mortality, including the length of time participants stayed in an intensive care unit, and in hospital.

The aim of the study was to evaluate whether the administration of convalescent plasma could decrease COVID-19 morbidity and mortality in patients with severe cancer. The most serious risks associated with the research were relatively rare and included the following:

1. Transfusion Related Acute Lung Injury (TRALI). Although a known risk in the context of COVID-19, TRALI is very difficult to identify since, in most cases when suspecting this complication, the patient is already in acute respiratory failure with damaged lungs.
2. Allergic reactions and anaphylaxis, which were estimated to be between 1 and 3 percent. Most of these reactions would be of a mild and transitory nature and would not require the transfusion to be suspended.

3. Transfusion-related volume overload. This is an infrequent condition, which the researchers proposed to manage and prevent by administering only 200 ml of convalescent plasma, at a slow rate of infusion, per session.

Patients would be invited to give consent and required to sign an informed consent form before participating in this study. If prospective participants were not competent to give consent, signed written consent would be required from a legal representative or a family member. This would create significant challenges, as it was current standard practice to isolate people with COVID-19 and to require all their contacts to quarantine for 14 days. It would therefore be difficult for research staff to meet with family members in person.

Questions
1. In a pandemic, should a research ethics committee make exceptions to standard requirements for informed consent? If so, what might be the limits to these exceptions?
2. Where it may not be possible to meet prospective participants' family members or legal representatives to obtain signed written consent, are there other ethically acceptable ways of seeking and documenting consent for this type of research?
3. In settings where many families do not have access to digital technologies, is it appropriate to request that participants' family members or legal representatives provide signed consent via email? Why?
4. What responsibilities should researchers have to evaluate and respond to potential socio-economic barriers to research participation?

References

Mair-Jenkins, J., M. Saavedra-Campos, J.K. Baillie, P. Cleary, F.M. Khaw, W.S. Lim, S. Makki, K.D. Rooney, J.S. Nguyen-Van-Tam, and C.R. Beck (Convalescent Plasma Study Group). 2015. The effectiveness of convalescent plasma and hyperimmune immunoglobulin for the treatment of severe acute respiratory infections of viral etiology: A systematic review and exploratory meta-analysis. *Journal of Infectious Diseases* 211(1): 80–90. https://doi.org/10.1093/infdis/jiu396.
Mei, H., and Y. Hu. 2020. Characteristics, causes, diagnosis and treatment of coagulation dysfunction in patients with COVID-19. *Zhonghua Xue Ye Xue Za Zhi* 41(3): 185–191. Chinese. https://doi.org/10.3760/cma.j.issn.0253-2727.2020.0002.
Mupapa, K., M. Massamba, K. Kibadi, K. Kuvula, A. Bwaka, M. Kipasa, R. Colebunders, and J.J. Muyembe-Tamfum. 1999. Treatment of Ebola hemorrhagic fever with blood transfusions from convalescent patients. *Journal of Infectious Diseases* 179(Supplement_1): S18–S23.

Case 9.4: A Study Involving Minimally Invasive Tissue Sampling in Adults Who Died from COVID-19

This case study was written by members of the case study author group.

Keywords Consent; Researcher roles and responsibilities; Research design and adaption; Safety and participant protection; Researcher safety

The spread and severity of COVID-19 are influenced by complex factors that vary globally, demonstrating the importance of region-specific research to enhance prevention, management and treatment of the disease. However, the infectious nature of COVID-19, and the morbidity and mortality associated with it, as well as infection-control measures such as quarantine and restrictions on public gatherings, present ethical challenges for research practices, including consent processes and risk minimization strategies for the study team, the research participants and the community.

A study involving post-mortem minimally invasive tissue sampling to investigate the pathogenic processes of COVID-19 infections in the lungs among adults was being conducted at a tertiary hospital in Africa in order to identify therapeutic targets. The study was being conducted among adults who had died of COVID-19, non-COVID-19 lower respiratory infections and infectious non-pulmonary illnesses. Studies involving minimally invasive tissue sampling already present cultural acceptability challenges in this setting, owing to fears of organ harvesting and disfigurement of the body. These challenges were compounded by other problems, including misinformation about COVID-19, the social stigma and discrimination faced by affected families, public apprehension over measures aimed at preventing COVID-19, and physical assault of COVID-19 health-care workers in some communities.

Communicating About the Study to Families

Considering the misinformation about COVID-19 present in the community, one of the study team's aims was to ensure the bereaved family or next of kin had enough information to make informed decisions. The study team could not inform patients or their next of kin about the study while they were in hospital because the patients were sick and family presence on the ward was strictly limited. The study team were also concerned that rumours circulating after previous studies could lead the next of kin to think that health-care workers had been negligent in the care of their family member in order to enrol them into the study after they died. The study team reflected on the safest, most culturally sensitive and ethically sound way of informing the next of kin about the study after the death of their family member. The idea of engaging a health-care worker who was both a study team member and a front-line worker in the hospital ward to notify the family about the study was considered to introduce a potential conflict of interest. As such, health-care workers who were not part of the study team were asked to notify the study team about a death and to seek consent from the next of kin to pass on their contact details.

After an initial phone call with the next of kin, the study team felt that a face-to-face meeting was important to enable a sensitive and sufficiently detailed discussion about the study, share the consent form and obtain informed consent. Traditionally in the study setting, decisions regarding the deceased involve many family members, as well as friends and sometimes community figures. The study team felt that visiting the house of the deceased for these discussions posed significant risks. The team also considered that assembling large groups of people at the hospital presented its own logistical challenges and carried infection risks. In the absence of clear ethical guidelines on how consent should be sought for post-mortem research in the context of the pandemic, the study team carefully weighed up cultural, logistical and safety factors that could determine which family members should give consent. These factors included patrilineal and matrilineal family systems; the age, gender and marital status of the deceased; and the availability of close family members, given that the study was conducted in an urban setting where most people had migrated for employment or business purposes. The legitimate next of kin to give consent was therefore determined by each family with guidance from the research team.

Risks to Family, Research Staff and the Community

Apart from the risks of physical harm to study team members if they visited the next of kin at their home, the study team also considered the public health risks of asking the next of kin to use public transport to travel to the hospital for the consent process. In addition, where the next of kin had been in close contact with the deceased, according to regulations they needed to be in quarantine. The study team developed a decision-making tree to cross-check all these details before inviting the next of kin to the facility. They also provided a study vehicle and personal protective equipment to minimize the risk of infection.

Questions
1. How should communication with patients and the next of kin about post-mortem research be adapted in the context of the pandemic?
2. How should consent processes for post-mortem research be adapted in the context of the pandemic? Can family consent be ethically waived in the context of the pandemic, given the greater good that could emerge from research on COVID-19 in the absence of proven treatment? Why?
3. What types of risks might study staff and next of kin face during discussions about this research in the context of the pandemic? How should these be minimized?

Case 9.5: COVID-19 Clinical Trials: Placebo Group Participants and the Right to Access the Experimental Product

This case study was written by members of the case study author group.

Keywords Access to experimental treatments; Resource allocation; Regulatory review; Vulnerability and inclusion; Risk/benefit analysis; Post trial follow up and monitoring; Placebo control; Vaccines

After the declaration of the COVID-19 pandemic in early 2020, scientific research accelerated with the objective of preventing and effectively treating COVID-19. The efforts of the global scientific community led to the development of more than 300 COVID-19 vaccine projects. In 2021 studies demonstrated that new vaccines may play an essential role in protecting individuals and reducing the spread of the virus (Forni and Mantovani 2021).

In a South American country, two double-blind multi-centre Phase III clinical trials with an experimental group and a control group got underway in 2020. In the studies a COVID-19 vaccine was given to the experimental group and a placebo to the control group. One of the clinical trials was coordinated by a public institution studying an inactivated SARS-CoV-2 vaccine, which aimed to stimulate the immune system to produce antibodies with a reduced risk of adverse events. The other clinical trial was coordinated by a pharmaceutical company and evaluated a live attenuated vaccine which incorporated a chimpanzee adenovirus.

The participants in these clinical trials were adult volunteers, with no history of infection with SARS-CoV-2, who provided care for patients with COVID-19. Exclusion criteria included pregnancy or plans to become pregnant within the next 3 months, specific illnesses, and health-care conditions requiring medications that would alter immune responses. In both Phase III trials the effectiveness of the experimental vaccines was demonstrated, which enabled emergency use authorization and approval by local regulatory agencies.

Following this approval, the government established a national COVID-19 immunization plan and requested that all doses produced by the two institutions be made available to the government for the immunization of the priority group in the initial phase of the campaign – front-line health professionals. This directly impacted the capacity of the two institutions to offer the vaccines to the volunteers in the placebo group of each trial. The volunteers in the clinical trials began receiving information about whether they had been in the experimental or control group 2 weeks after the national immunization plan started.

Questions
1. What are the ethical implications of the government's request for all of the vaccine doses produced by the two research institutions?
2. Since the research provided evidence on the efficacy and safety of two vaccines during the pandemic, should participants in the studies' control groups be prioritized to receive the vaccine over other priority groups? Why?

3. Once the vaccines have been shown to be efficacious and safe, should participants be informed about whether they were allocated to the control or experimental arm of the study? Why?
4. Are the inclusion and exclusion criteria for these studies appropriate and justifiable in the context of the COVID-19 pandemic? Why?

Reference

Forni, G., and A. Mantovani. 2021. COVID-19 vaccines: Where we stand and challenges ahead. *Cell Death & Differentiation* 28(2): 626–639. https://doi.org/10.1038/s41418-020-00720-9.

References

Appiah, R. 2021. Gurus and griots: Revisiting the research informed consent process in rural African contexts. *BMC Medical Ethics* 22(1): 1–11.

Brown, N. 2018. *Video-conference interviews: Ethical and methodological concerns in the context of health research.* SAGE. https://doi.org/10.4135/9781526441812.

Chiumento, A., L. Machin, A. Rahman, and L. Frith. 2018. Online interviewing with interpreters in humanitarian contexts. *International Journal of Qualitative Studies on Health and Well-Being* 13(1). https://doi.org/10.1080/17482631.2018.1444887.

CIOMS. 2016. *International ethical guidelines for health-related research involving humans.* Geneva: Council for International Organizations of Medical Sciences.

Fox, A., S. Baker, K. Charitonos, V. Jack, and B. Moser-Mercer. 2020. Ethics-in-practice in fragile contexts: Research in education for displaced persons, refugees and asylum seekers. *British Educational Research Journal* 46(4): 829–847. https://doi.org/10.1002/berj.3618.

Goldman, R.D., and L. Gelinas. 2021. COVID-19 and consent for research: Navigating during a global pandemic. *Clinical Ethics* 16(3): 222–227.

Goyal, M., J.M. Ospel, A. Ganesh, M. Marko, and M. Fisher. 2021. Rethinking consent for stroke trials in time-sensitive situations: Insights from the COVID-19 pandemic. *Stroke* 52(4): 1527–1531.

Hofmann, B. 2020. The first casualty of an epidemic is evidence. *Journal of Evaluation in Clinical Practice* 26: 1344–1346. https://doi.org/10.1111/jep.13443.

Huang, B., and K.M. Carley. 2020. Disinformation and misinformation on Twitter during the novel coronavirus outbreak. *arXiv* Preprint arXiv:2006.04278. https://arxiv.org/abs/2006.04278.

Karnieli-Miller, O., R. Strier, and L. Pessach. 2009. Power relations in qualitative research. *Qualitative Health Research* 19(2): 279–289. https://doi.org/10.1177/1049732308329306.

Kass, N.E., J. Sugarman, R. Faden, and M. Schoch-Spana. 1996. Trust, the fragile foundation of contemporary biomedical research. *Hastings Center Report* 26(5): 25–29.

Kerasidou, A. 2017. Trust me, I'm a researcher! The role of trust in biomedical research. *Medicine, Health Care and Philosophy* 20(1): 43–50. https://doi.org/10.1007/s11019-016-9721-6.

Kingori, P. 2015. The "empty choice": A sociological examination of choosing medical research participation in resource-limited sub-Saharan Africa. *Current Sociology* 63(5): 763–778.

Lairumbi, G.M., M. Parker, R. Fitzpatrick, and M.C. English. 2011. Ethics in practice: The state of the debate on promoting the social value of global health research in resource poor settings particularly Africa. *BMC Medical Ethics* 12(1): 1–8.

———. 2012. Forms of benefit sharing in global health research undertaken in resource poor settings: A qualitative study of stakeholders' views in Kenya. *Philosophy, Ethics, and Humanities in Medicine* 7(1): 1–8.

Largent, E.A., S.D. Halpern, and H. Fernandez Lynch. 2021. Waivers and alterations of research informed consent during the COVID-19 pandemic. *Annals of Internal Medicine* 174(3): 415–416. https://doi.org/10.7326/M20-6993.

Lo Iacono, V., P. Symonds, and D.H.K. Brown. 2016. Skype as a tool for qualitative research interviews. *Sociological Research Online* 21(2): 103–117. https://doi.org/10.5153/sro.3952.

Macioce, F. 2021. Informed consent and group vulnerability in the context of the pandemic. *BioLaw Journal – Rivista di BioDiritto* 2S: 17–33. https://doi.org/10.15168/2284-4503-844.

Miller, P.B., and C. Weijer. 2006. Trust based obligations of the state and physician-researchers to patient-subjects. *Journal of Medical Ethics* 32(9): 542–547.

Nuffield Council on Bioethics. 2020. *Research in global health emergencies: Ethical issues.* London: Nuffield Council on Bioethics.

O'Neill, O. 2002. Autonomy and trust in bioethics. In *The concept of human dignity in biomedical law*, ed. M. Geier and P. Schröder. Cambridge: Cambridge University Press.

Resnik, D.B. 2011. Scientific research and the public trust. *Science and Engineering Ethics* 17(3): 399–409. https://doi.org/10.1007/s11948-010-9210-x.

Whitbeck, C. 1995. Truth and trustworthiness in research. *Science and Engineering Ethics* 1(4): 403–416.

WHO. 2020. *Ethical standards for research during public health emergencies: Distilling existing guidance to support COVID-19 R&D*. Geneva: World Health Organization. https://www.who. int/blueprint/priority-diseases/key-action/liverecovery-save-of-ethical-standards-for-research- during-public-health-emergencies.pdf.

Wilkinson, D., J. Savulescu, T. Hope, and J. Hendrick. 2008. *Medical ethics and law: The core curriculum.* Elsevier Health Sciences.

Chapter 10
Afterword

Susan Bull ⓘ and Michael Parker

Abstract This casebook offers a window into important aspects of the ethical landscapes that researchers, communities, health professionals, policy makers – and ethicists – had to navigate during the first 15 months of the COVID-19 pandemic. The cases presented in this casebook are inevitably a selection informed by and constrained by the processes through which they were sought, and by the pandemic itself. Additional cases could valuably complement all the thematic chapters in this casebook. In addition, this casebook calls for a broader approach to research ethics, both in terms of the issues to be considered, and the range of stakeholders having ethical responsibilities relating to the conduct of research. However a broad range of stakeholders have differing values, remits, authorities and capacities to exercise power in pandemic contexts, and in many situations, exercises of power, and their impact on research, are not direct and explicit. As such they are less amenable to clear representation in real-world cases, highlighting the importance of complementing discussions of the cases in this casebook with conceptual literature. Reflection on the research that has not been conducted is also critical. The COVID-19 pandemic has reemphasized that global health emergencies are never only about health. The wide-ranging impacts of the pandemic on economies, employment, education and a range of socially and culturally important activities, accentuates the importance of an equally comprehensive research agenda, which goes beyond a narrow conception of 'health', and addresses a broad range of pandemic impacts on populations. A further way in which we believe debate on pandemic research ethics both could and should be broadened is in relation to aspects of pandemic science beyond those relating to 'response'. Inevitably, in the

S. Bull (✉)
The Ethox Centre, Nuffield Department of Population Health, University of Oxford, Oxford, Oxfordshire, UK

Faculty of Medical and Health Sciences, University of Auckland, Auckland, New Zealand
e-mail: susan.bull@ethox.ox.ac.uk

M. Parker
The Ethox Centre, Nuffield Department of Population Health, University of Oxford, Oxford, Oxfordshire, UK

© PAHO and Editors 2024
S. Bull et al. (eds.), *Research Ethics in Epidemics and Pandemics: A Casebook*,
Public Health Ethics Analysis 8, https://doi.org/10.1007/978-3-031-41804-4_10

context of an emerging and continuing pandemic, scientific research attention has tended to focus on interventions that can enable more effective responses. However pandemic science can be thought of as divisible into four interdependent and overlapping domains: prevention, preparedness, response, and recovery. Research is essential to the development, evaluation, and deployment of interventions in each of these domains and effective, valuable, trustworthy and trusted research will require ethical questions to be identified and addressed. This chapter concludes by inviting the connection of additional cases and conceptual resources to this case-book, to enhance and expand the themes and topics covered.

Keywords COVID-19 pandemic · Research ethics · Public health emergency response · Public health emergency preparedness · Public health emergency prevention · Public health emergency recovery

10.1 Introduction

The COVID-19 pandemic has had – and continues to have - an unprecedented global and 'whole of society' impact (Miyah et al. 2022; Elavarasan et al. 2022). The critical ethical importance of conducting timely, rigorous and responsive research to inform public health and clinical responses to the pandemic, and to ensure that lessons are learned to inform responses to future emergencies, is inarguable (Swaminathan et al. 2022). So too is the evidence that pandemics raise a number of distinctive and profound challenges for the conduct of such research. Health emergencies are contexts in which effective and informative research conducive to well-founded public trust and confidence is desperately needed. They are also, however, almost by definition situations that are radically non-ideal for such research (Nuffield Council on Bioethics 2020).

Many of the challenges arising when conducting research in emergencies have an ethical dimension. From the very beginning of the COVID-19 pandemic, there has been a complex, rapidly evolving and at times contested literature addressing a range of ethical issues arising in pandemic health research. A comprehensive list is not feasible, but examples of areas in which such issues have been discussed include: decisions about prioritising, funding, suspending and halting biomedical and social science research; accelerated research and vaccine development pathways; adapting and adaptive research; research exceptionalism, quality and misconduct; monitored emergency use of unregistered and investigational interventions; adapted and expe-dited ethics review pathways, and adaptions to governance and oversight mecha-nisms for national and multi-national studies; unprecedented rates of scientific pre-publication, publication and retraction; participant protection and inclusive approaches to recruitment; curation, analysis and sharing of phylogenetic, surveil-lance and research data; and research using mobile apps and social media data. This literature has made an important contribution to public and academic debate and to the development of policy. However, it has, at times, been inadequately informed by an in-depth understanding of practical ethical issues arising on the ground in research contexts around the world. To address this deficit, in December 2020 we issued an open global call to researchers, reviewers and academics to contribute case-studies and co-create a capacity building resource informed by, and responsive to, lived experiences. The 44 cases within this casebook, drawn from Africa, Asia, the Americas, Europe and Oceania, provide contextually rich examples of practical ethical issues arising as health-related research was conducted during 2020 and early 2021. While this means that casebook provides a critical insight into lived experiences and complex challenges on the ground from a broad range of settings, it is not possible for such a compilation to be exhaustive. This is true in at least two important respects – it not possible to represent all the contexts and settings in which research has taken place, nor to investigate all the ethical dimensions of the multiplicity of biomedical and social science research approaches undertaken in the context of the COVID-19 pandemic.

Having said this, the cases in this casebook are not intended to capture all ethical dimensions of challenges arising when conducting research, nor are they necessarily intended to be representative. Our aim here has been much more limited. We have sought to compile a diverse and inclusive range of cases and thematic commentaries with the potential to be used to prompt discussion and reflection on the ethical issues presented both by these specific cases and beyond them. It is our hope that readers will make links between the cases discussed here and those challenges arising in their own experience or elsewhere and reflect upon the relevance, and the limitations, of the commentaries presented here for those situations.

10.2 Looking Forward

It is important to reflect both on areas that have received less attention in this casebook, and to emphasise the importance of conducting further work into the broad range of important ethical issues arising during research conducted in the differing contexts of global health emergencies. An initial consideration is that for many of the themes addressed in the casebook, *additional cases could provide an important resource to prompt more nuanced consideration of relevant ethical dimensions.* For example, the collection and analysis of case studies of research with migrant or refugee populations would form a valuable complement to the discussions of vulnerability, inclusion and protections in Chaps. 8 and 9, and prompt discussion of further relevant ethical issues, including those associated with roles and responsibilities of community 'gatekeepers' in emergency contexts. Many other examples could be given and we would love to see this resource connected to others that enhance and expand the themes and topics covered.

In addition, the breadth, complexity, and at times inter-relatedness of the ethical considerations arising in the majority of cases prompt (and are intended to prompt) consideration of a wider range of ethical issues than those associated with traditional accounts of research ethics. Chapter 1 calls for readers to take a broader approach to research ethics – both in terms of the issues to be considered, and the actors having ethical responsibilities relating to the conduct of research is pandemics and other health emergencies. Examples include the nature of responsibilities to address systemic inequities in capacity to rapidly pivot and originate and conduct responsive research, and rollout effective interventions in pandemics. Important questions also arise about the supra-research roles and responsibilities of research organisations in pandemic contexts, including for example, supporting national COVID-19 testing and pathogen sequencing initiatives and the provision of healthcare.

Global health is inherently a complex field of social, political, economic, and scientific relationships in which a broad range of stakeholders have differing values, remits, authorities and capacities to exercise power. Issues relating to the exercise of epistemic power are important to consider, as this is a space within which questions of expertise, what constitute reliable sources of information and evidence, and the role and responsibilities of the media, are inevitably contested. Within complex global health landscapes, the COVID-19 pandemic has both emphasised and exacerbated global health and epistemic inequities, demonstrating the importance of critically reflecting on the roles and responsibilities of powerful actors in international and national pandemic research and response. Such stakeholders are sometimes far from the ethical issues experienced by researchers and reviewers, but their decisions profoundly influence which research is conducted and the options available to researchers and the communities in which they work.

While the cases in Chap. 1 directly address the consequences of decision-making by national and multinational stakeholders, *in many situations, exercises of power, and their impact on research, are less direct and explicit, and are thus less amenable to clear representation in real-world cases.* In the context of the COVID-19 pandemic for example, unprecedented political, economic and social pressures to develop and implement effective responses have had significant implications for accelerated vaccine development pathways. National and multinational interests have profoundly influenced processes for emergency use authorisations of COVID-19 vaccines, including the design and implementation of COVID-19 vaccine studies, the fairness of research collaborations, and approaches to pre-publication and the sharing (or not) of research data. Such interests also had profound implications for both enablers and barriers to the collective action required for post-trial access to vaccines, and global vaccine equity. These and other exercises of power highlight the importance of complementing discussions of the cases in this casebook with an appreciation of the relevant conceptual literatures focusing on the responsibilities of a range of stakeholders when research is conducted in pandemics and other health emergencies.

The COVID-19 pandemic has also reemphasized that global health emergencies are never only about health. The wide-ranging impacts of the pandemic on economies, employment, education and a range of socially and culturally important activities, accentuates the importance of an equally comprehensive research agenda, which goes beyond a narrow conception of 'health', and addresses a broad range of pandemic impacts on populations. While the cases in this casebook promote consideration of the ethical issues arising in the research conducted during the first 15 months of the pandemic, *reflection on the research that had not been conducted is also vital.* As discussed in Chap. 2, further work is needed to explore what a comprehensive research agenda in a public health emergency should comprise, the values that should inform such decision-making, and inclusive and procedurally fair approaches to developing research that is responsive to population burdens. It is important to explore effective approaches to developing research priorities which leave no-one behind in pandemic contexts where systemic inequalities and inequities are exacerbated and have broad ranging impacts both on health, and on social determinants of heath. This includes exploring the impacts of pausing and halting research into pre-pandemic health priorities, and the impacts such revisions have had, for example, on the research populations involved, and on development of evidence to inform effective approaches to addressing their health needs.

During the COVID-19 pandemic, in addition to substantive developments in research pathways for the development of diagnostics, vaccines, and evaluations of novel and repurposed therapeutics, there has been unprecedented implementation of non-clinical approaches to reducing the transmission of infection, sometimes referred to as 'non-pharmaceutical interventions' (NPIs). These have ranged from mandatory measures such as national lockdowns, vaccine passports, and school closures, through to interventions such as public information campaigns, the uses of 'influencers', and calls for 'solidarity'. The dramatic impacts of approaches to restrict population movement, and requirements for isolation, social distancing and mask-wearing, demonstrate that vitally important research during pandemics includes social science and public health research on the lessons to be learned from the differing approaches to the rollout and removal of mandated NPIs around the world. These include questions about how trade-offs were made between the importance of averting harms of COVID-19 and the burdens of such approaches. Important questions arise about the nature and magnitude of economic, educational, social, psychological and health burdens associated with NPIs, the distribution of such burdens amongst populations, and the effectiveness of approaches to equitably maximising the benefits of NPIs and ameliorating burdens (Osterrieder et al. 2021; Schneiders et al. 2022). Examples include the impacts of closing schools, and requirements for social distancing and mask-wearing, on education and social development in young children; the impacts of limiting access to social support and health services for vulnerable populations; and the impacts of requirements to self-isolate or enter quarantine on employment, housing, food-security and well-being (Phuong et al. Submitted). All these kinds of research present ethical challenges when undertaken in the context of a pandemic, which are compounded in contexts that are already subject to significant disadvantage. Careful consideration is also needed of the ethical issues that can arise when conducting research into health policies, including, for example, the ethics of conducting cluster-randomized trials to evaluate consequences of relaxing restrictions, and permitting large scale social events to take place as 'research activities' to explore effects on COVID-19 transmission rates. Moreover, in addition to the issues relating to surveillance and secondary uses of data addressed in Chap. 7, further work is needed to address the impacts of, and issues arising during, the rapid development and implementation of novel digital surveillance strategies to track and trace populations. These include issues relating to equitable development and rollout of such technologies and compliance with social mandates for uses of surveillance data.

Chapters 3, 4 and 5 additionally highlight the importance of undertaking further work to determine the advice and support needed to strengthen the capacities of researchers, regulators, ethical reviewers and health authorities to appropriately respond to incomplete, rapidly evolving, and at times problematic and contested research findings and pandemic evidence landscapes. In doing so, it is important to recognise that such landscapes, and accompanying infodemics, impact research agendas and prioritisation, and demonstrate the importance of conducting research into approaches to develop trusted and trustworthy engagement with populations in public health emergencies. (Borges do Nascimento et al. 2022) Such engagement is critical to meet responsibilities to promote public understanding of complex and evolving pandemic contexts, enable populations to take effective measures to promote their health and wellbeing, and mobilise the collective action approaches which are key to implementing effective whole of society responses to the whole of society impacts of pandemics. As discussed in Chap. 6, the COVID-19 pandemic also prompts further research into the practical ethical issues arising when implementing approaches to rapidly accelerating and adapting research review and oversight processes, in ways which seek to maintain substantive requirements to protect participants appropriately. Further work is also needed to explore stakeholders' roles and responsibilities to govern and co-ordinate research efforts in contexts where multiple small scale and potentially poor quality studies may be developed in response to perceived pressures to 'do something' to alleviate pandemic burdens, and where the purported social value of addressing pandemic burdens can be an exceptional multiplier in analyses of the potential benefits of research.

10.3   Beyond Pandemic Response

A further way in which we believe debate on pandemic research ethics both could and should be broadened is in relation to aspects of pandemic science beyond those relating to 'response'. Inevitably, in the context of an emerging and continuing pandemic, scientific research attention has tended to focus on interventions that can enable more effective responses. As illustrated in the cases presented in this casebook, this can include scientific research on the development and evaluation of diagnostics, vaccines, and therapeutics, and social science research on the impacts of NPIs. This tendency to focus on response is reflected in the cases collected for this casebook, which are concerned primarily with the ethics of research of this type. However, despite the importance and the urgency of developing and implementing effective responses, it is clear that pandemic science is or ought to be concerned with questions beyond those related to effective pandemic response. Indeed, pandemic science can be thought of as divisible into four interdependent and overlapping domains: prevention, preparedness, response, and recovery. Research is essential to the development, evaluation, and deployment of interventions in each of these four domains. In each of these domains effective, valuable, trustworthy and trusted research will require the identification, analysis, and addressing of ethical questions.

In the context of pandemic prevention, for example, research will need to be undertaken to inform the effective uses of surveillance and the analyses of data deriving from surveillance strategies. These data are likely to include information of a range of different kinds: genomic, syndromic, social media, satellite and drone imagery, in addition to clinical records and other sources. Research on prevention is also needed to investigate the effects of climate change, environmental degradation, and farming practices on the emergence of infectious diseases with outbreak, epidemic, and pandemic potential.

With regard to pandemic 'preparedness' research will also be crucial. Examples of research may include research related to early-stage vaccine platform development, and research about the factors underpinning health system resilience, amongst other priorities. As with pandemic prevention, however, much of the research relating to pandemic preparedness is necessarily going to focus on surveillance and the collection, curation, sharing, and analysis of data. A wide range of ethical questions arise and need to be addressed in the context of pandemic preparedness research, and a ground-up case-based approach such as that used this casebook would be an ideal way to do this. Some of these ethical questions will concern structural issues, such as those relating to global health inequities, in which case-based approaches will need to be complemented by capacity-strengthening resources of other kinds, as discussed above. However, contextually-rich cases about the implications of the different manifestations of global health injustice on the ground can play a key role in enhancing and informing discussions and debates about appropriate responses, and responsibilities, to address such inequities.

Finally, the 'recovery' phase of pandemics is also one in which medical, scientific, and social science research is both needed and will raise important ethical questions (British Academy 2021). Some of these will concern the ethical implications of the very decisions to declare emergencies 'over', including the implications such decisions have for the populations most directly affected by such emergencies, and for stakeholders' perceptions of their responsibilities to ameliorate ongoing burdens and impacts (Wadman 2022). Additional ethical questions will arise in the context of responsibilities to conduct research into the complex and multi-faceted longer-term effects of emergencies, outbreaks, epidemics and pandemics on populations. Taken together these suggest that there is value in expanding the use of a case based approach to include the ethical issues in research beyond questions relating to response and to include those arising in research undertaken as part of epidemic and pandemic prevention, preparedness and recovery.

10.4   Final Thoughts

In this afterword, we have discussed the ways in which the cases presented in this casebook are inevitably a selection, and moreover a selection both informed by and constrained by the processes through which they were sought and developed, and by the pandemic period through which we have all lived. These complex real-world cases offer a window into some important aspects of the ethical landscapes that researchers, communities, health professionals, policy makers – and ethicists – have had to navigate during the COVID-19 pandemic. Despite the fact that this collection of cases cannot be comprehensive, we believe that they provide an important resource for promoting and facilitating reflection on the ethics of research conducted during health emergencies. We have discussed the breadth and scope of research required in the context of the pandemic, and welcome this resource being connected to additional cases and conceptual resources that enhance and expand the themes and topics covered.

References

British Academy. 2021. *Shaping the Covid decade: Addressing the long-term societal impacts of COVID-19*. London: British Academy. https://www.thebritishacademy.ac.uk/documents/3239/ Shaping-COVID-decade-addressing-long-term-societal-impacts-COVID-19.pdf.

Borges do Nascimento, I.J., A.B. Pizarro, J.M. Almeida, N. Azzopardi-Muscat, M.A. Gonçalves, M. Björklund, and D. Novillo-Ortiz. 2022. Infodemics and health misinformation: A systematic review of reviews. *Bulletin of the World Health Organization* 100(9): 544–561. https://doi.org/ 10.2471/BLT.21.287654.

Elavarasan, R.M., R. Pugazhendhi, G.M. Shafiullah, et al. 2022. Impacts of COVID-19 on sustainable development goals and effective approaches to maneuver them in the post-pandemic environment. *Environmental Science and Pollution Research* 29: 33957–33987. https://doi.org/ 10.1007/s11356-021-17793-9.

Miyah, Y., M. Benjelloun, S. Lairini, and A. Lahrichi. 2022. COVID-19 impact on public health, environment, human psychology, global Socioeconomy, and education. *Scientific World Journal*: 5578284. https://doi.org/10.1155/2022/5578284.

Nuffield Council on Bioethics. 2020. *Research in global health emergencies: Ethical issues*. London: Nuffield Council on Bioethics. https://www.nuffieldbioethics.org/topics/research-ethics/research-in-global-health-emergencies.

Osterrieder, A., G. Cuman, W. Pan-Ngum, et al. 2021. Economic and social impacts of COVID-19 and public health measures: Results from an anonymous online survey in Thailand, Malaysia, the UK, Italy and Slovenia. *BMJ Open* 11: e046863. https://doi.org/10.1136/bmjopen-2020-046863.

Phuong, T.T., M. Duwal, D. Timoria, I.A. Sutrisni, S. Rijal, C. Bogh, A. Kekalih, D. Friska, A. Karkey, R.L. Hamers, S. Lewycka, M. Chambers, on behalf of the OUCRU COVID-19 Research Group, and J.I. Van Nuil (Submitted) Exploring layers of vulnerability during COVID-19: Qualitative research with 13 communities in Indonesia, Nepal, and Vietnam.

Schneiders, M.L., B. Naemiratch, P.K. Cheah, G. Cuman, T. Poomchaichote, S. Ruangkajorn, et al. 2022. The impact of COVID-19 non-pharmaceutical interventions on the lived experiences of people living in Thailand, Malaysia, Italy and the United Kingdom: A cross-country qualitative study. *PLoS One* 17(1): e0262421. https://doi.org/10.1371/journal.pone.0262421.

Swaminathan, S., B. Pécoul, H. Abdullah, C. Christou, G. Gray, C. Ijsselmuiden, M.P. Kieny, et al. 2022. Reboot biomedical R&D in the global public interest. *Nature* 7896: 207–210. https://doi.org/10.1038/d41586-022-00324-y.

Wadman, M. 2022. When is a pandemic 'over'?: World Health Organization prepares to confront thorny decision. *Science* 375(6585): 1077–1078. https://doi.org/10.1126/science.adb1919.

Correction to: Research Quality and Dissemination

Sergio Litewka and Sarah Sullivan

Correction to:
Chapter 3 in: S. Bull et al. (eds.), *Research Ethics in Epidemics and Pandemics: A Casebook*, **Public Health Ethics Analysis 8, https://doi.org/10.1007/978-3-031-41804-4_3**

The chapter was inadvertently published with home address of co-author Sarah Sullivan instead of the institutional affiliation.

This error has been corrected by updating the institutional affiliation "College of Education and Health Sciences, Touro University California, Vallejo, CA, USA".

The updated version of this chapter can be found at
https://doi.org/10.1007/978-3-031-41804-4_3

© PAHO and Editors 2024
S. Bull et al. (eds.), *Research Ethics in Epidemics and Pandemics: A Casebook*,
Public Health Ethics Analysis 8, https://doi.org/10.1007/978-3-031-41804-4_11

Index